Table of Contents

Chapter 1

Learning Objectives
- Understand data
- Identify the scale of measurement of a variable
- Determine if a variable is quantitative or qualitative, discrete or continuous

Defining Data

Data are observations (measurements, survey responses, etc) that have been collected. Data are usually just a set of numbers. They could be a group of numbers representing heights, for example. We call this group of numbers a variable because with each subject, the number changes. So, from person to person the numbers vary.

Some data sets consist of <u>numbers</u> (*height*) and others are <u>nonnumeric</u> (*eye color*). For example, the labels of male and female are values for the variable gender, and this would be nonnumeric data. When we enter data in SPSS, we will make this data numeric by assigning number values to the labels of male and female, like 1 is the value for a male and 2 is the value for a female.

We often describe numeric data as Quantitative and nonnumeric data as Qualitative.

1. <u>Quantitative data</u> – numbers representing counts or measurements.
   a. Example - Incomes of college graduates.
2. <u>Qualitative (or categorical) data</u> – can be separated into different categories that are distinguished by some nonnumeric characteristic.
   a. Example - Genders of college graduates – male or female.

We can further describe *QUANTITATIVE* data as follows:
1. <u>Discrete</u> – possible values are a finite or "countable" number. There are no possible values in between each number.
   a. Examples
      i. – number of eggs that hens lay (0,1,2, etc.)
      ii. – number of phone calls received per day (You could not receive 6.3 calls per day, that doesn't make sense.)
2. Continuous – infinitely many possible values, a continuous scale.
   a. Example
      i. - Amount of milk a cow produces. (Any value between 0-5, like 2.3412 gallons)

ii. – Weight of a person (153.543 pounds) Note that your weight could take on infinitely many values, based on the number of decimal points you include.

Level of Measurement

What we will focus on is the level of measurement used. This is the first and most important distinction we must make about a variable before doing statistical analysis! Statisticians often refer to the levels of measurement of a variable, which is a scale to distinguish between measured variables that have different properties. There are 4 levels of measurement in ascending order of precision are: nominal, ordinal, interval, and ratio.

Different types are measured differently. For instance, to measure time to run the 50 meter dash, you might use a stop watch, and report the time in seconds. This would be a ratio variable. To measure someone's attitude toward a political candidate, a rating scale with labels like "very favorable", "somewhat favorable", etc., would be more useful than a stop watch. This type of variable is ordinal. For a variable like favorite color, we would simply note the color word, like red that the subject offers. This is a nominal variable.

Another Example: Party Affiliation

Assume only relevant attributes are: republican, democrat, and independent.

To analyze the results of this variable, we arbitrarily assign the values 1, 2 and 3 to the three attributes. The *level of measurement* describes the relationship among these three values.

In this case, we are using the numbers as shorter placeholders for the lengthier text terms. We don't assume that higher values mean "more" of something and lower numbers signify "less". We would describe the level of measurement as "nominal".

Four levels of measurement

I. NOMINAL

At the first level of measurement, we can use numbers to represent words or letters for the responses. In nominal measurement the numerical values just "name" the attribute uniquely. The data cannot be arranged in an ordering scheme. One value is really not any greater than another. A good example of a nominal variable is sex (or gender). Information in a data set on sex is usually coded as 0 or 1, 1 indicating male and 0 indicating female (or the other way around--0 for male, 1 for female). 1

in this case is an arbitrary value and it is not any greater or better than 0. There is only a nominal difference between 0 and 1. With nominal variables, there is a qualitative difference between values, not a quantitative one.

If you wanted to classify a football team into left footed and right footed players, you could put all the left footed players into a group classified as 1 and all the right footed players into a group classified as 2. The numbers 1 and 2 are used for convenience; you could equally use the letters L and R, or the words LEFT and RIGHT to label the groups of players. Numbers are often preferred because text takes longer to type out and takes up more space.

Another example of a nominal variable is blood groups where the letter A, B, O and AB represent the different classes.

More Examples:

1. **Variable=jersey number** A player with number 30 is not more of anything than a player with number 15, and is certainly not twice whatever number 15 is.

2. **Variable = SSN** Social security numbers are for identification only. They don't represent measurements or counts of anything.

3. Survey responses of yes, no, undecided.

Other examples of nominal variables: zodiac sign, gender, ethnicity, and religious affiliation

II. ORDINAL

In **ordinal** measurement the attributes can be rank-ordered, but differences between values are meaningless.

Examples:

1. **Variable= Educational Attainment** 0=less than H.S.; 1=some H.S.; 2=H.S. degree; 3=some college; 4=college degree; 5=post college. In this measure, higher numbers mean *more* education. But is distance from 0 to 1 same as 3 to 4? Of course not. The interval between values is not interpretable in an ordinal measure.
2. **Variable=Car size** Compact, Mid-size, Full-size. Full-size is larger than compact, but no exact difference can be found. You can't subtract compact from full-size and get a meaningful result. We do not know the magnitude of the difference between full-size and compact cars.
3. **Variable=Social Class** Upper, Middle, Working class.
4. We could order the finishers in a race or the students by their grade point average (first in class, second in class, and so forth down to last in class). Notice that the intervals between cases are probably not the same. The class valedictorian may have a 4.0, the salutatorian a 3.6, and the third student a 3.5.
5. Likert scales – strongly agree, agree, disagree, strongly disagree are ordinal variables. We can't define how much more intensely a person agrees or disagrees with a statement.

3

III. INTERVAL

In **interval** measurement the distance between attributes *does* have meaning. Interval variables possess a common and equal unit that separates adjacent categories. However, there is no natural zero starting point (there is no value where none of the quantity is present).

Examples:

1. **Variable=temperature (Fahrenheit)** The distance from 30-40 is same as distance from 70-80. The interval between values is interpretable. Because of this, it makes sense to compute an average of an interval variable, where it doesn't make sense to do so for ordinal scales. However, there is no natural starting point. The value of 0 degrees (F) is arbitrary and does not represent the total absence of heat.
2. **Variable=Year** Years like 1000, 1776, 2002. (Time did not begin in the year 0, so year 0 is arbitrary instead of being a natural zero starting point).

IV. RATIO

Finally, in **ratio** measurement there is always an absolute zero that is meaningful. This means you can't go below a score of zero for this variable. This means that you can construct a meaningful fraction (or ratio) with a ratio variable.

Examples:

1. Weight
2. Length
3. Age
4. Income
5. In applied social research most "count" variables are ratio, for example, the number of clients in past six months. Why? Because you can have zero clients and because it is meaningful to say that "...we had twice as many clients in the past six months as we did in the previous six months."
6. Student commuting distances - 5 miles, 20 miles, 40 miles. It is meaningful to say 40 miles is twice as far as 20 miles, and 0 is a natural starting point.

There is a hierarchy implied in the level of measurement. At lower levels of measurement, assumptions tend to be less restrictive and data analyses tend to be less sensitive. At each level up the hierarchy, the current level includes all of the qualities of the one below it and adds something new. In general, it is desirable to have a higher level of measurement (e.g., interval or ratio) rather than a lower one (nominal or ordinal).

In SPSS, there are 3 levels of measurement: Nominal, Ordinal, and Scale. Scale data values are numeric values on an interval or ratio scale (e.g., age, income). Scale variables must be numeric.

Again, ordinal data values represent categories with some intrinsic order (e.g., low, medium, high; strongly agree, agree, disagree, strongly disagree). Ordinal variables can be either string (alphanumeric) or numeric values that represent distinct categories (e.g., 1=low, 2=medium, 3=high). Note: for ordinal string variables, the alphabetic order of string values is assumed to reflect the true order of the categories. For example, for a string variable with the values of low, medium, high, the order of the categories is interpreted as high, low, medium -- **which is not the correct order**. In general, it is more reliable to use numeric codes to represent ordinal data. (i.e., 1=low, 2=medium, 3=high).

Nominal. Data values represent categories with no intrinsic order (e.g., job category or company division). Nominal variables can be either string (alphanumeric) or numeric values that represent distinct categories (e.g., 1=Male, 2=Female).

Why is Level of Measurement Important (i.e. Who Cares)?

1. Interpreting data from that variable. *Do the numerical codes mean something or are they just codes, as in the party affiliation variable?*
2. Knowing the level of measurement helps you decide what statistical analysis is appropriate or what statistics can be meaningfully calculated with that variable. *If a measure is nominal, you would never average the data values.*

For example, consider a hypothetical study in which 5 children are asked to choose their favorite color from blue, red, yellow, green, and purple. The researcher codes the results as follows:

Color	Code
Blue	1
Red	2
Yellow	3
Green	4
Purple	5

This means that if a child said her favorite color was "Red," then the choice was coded as "2," if the child said her favorite color was "Purple" then the response was coded as 5, and so forth.

Consider the following hypothetical data:

Subject	Color	Code
1	Blue	1
2	Blue	1
3	Green	4
4	Purple	5
5	Green	4

Each code is a number, so nothing prevents us from computing the average code assigned to the children. The average happens to be 3, but you can see that it would be senseless to conclude that the average favorite color is yellow (the color with a code of 3).

PRACTICE EXERCISES

Answer the following questions by checking the appropriate response.

1. In an experiment to determine if a specific feed increases the final weight of cattle, the following were measured on each animal in the study.

> Gender, Initial weight, Weight gain, Grade of meat, where grade or meat is recorded as (A, B, or C).

> The scale of measurement of these variable (in order) are:
> (a) **Nominal**, scale, scale, **nominal**
> (b) **Nominal**, scale, scale, **ordinal**
> (c) **Ordinal**, scale, scale, **ordinal**
> (d) **Ordinal**, scale, scale, **nominal**

2.   A researcher measures juvenile delinquency by asking juveniles how they feel about smoking pot. They choose one of the following answers:

> People who smoke pot are good. I strongly agree with it.
> Smoking pot is not a big deal. I agree with it.
> I don't care if people smoke or not. I'm neutral.
> Smoking pot can get you in trouble. I don't agree with it.
> People who smoke pot are losers. I strongly disagree with it!

NOMINAL   ORDINAL   INTERVAL/RATIO (SCALE)

3. A researcher decides to collapse the social class [income] variable into the following categories:

> 0 = $0-$10,000
> 1 = $10,001-$20,000
> 2 = $20,001-$30,000
> 3 = $30,001-$40,000
> 4 = $40,001-$50,000
> 5 = $50,001+

NOMINAL   ORDINAL   INTERVAL/RATIO (SCALE)

4. A study was conducted to investigate the effect of a nuclear plant in Ohio on the water quality of the Ohio river. As part of an environmental impact study, fish were captured, tagged, and released. The following information was recorded for each fish:

Gender (1=female, 2=male), Length (inches),Maturation (0=young, 1=adult),Weight (oz).

The **scale** of these variables is:

(a) **nominal**, ratio, **nominal**, ratio
(b) **nominal**, interval, **ordinal**, ratio
(c) **nominal**, ratio, **ordinal**, ratio
(d) **ordinal**, ratio, **nominal**, ratio
(e) **ordinal**, interval, **ordinal**, ratio

5. Starting in the West and heading counter clockwise around the country, a researcher labels Geographical location as
0 = West 1 = Midwest 2 = East 3 = South

NOMINAL ORDINAL INTERVAL/RATIO (SCALE)

6. The following variables were measured at several sites around Iowa:

pH of soil (to one decimal place, e.g., 6.3);
crop grown (0=corn, 1=barley, 2=oats, 3=other);
amount of stubble (0=light, 1=medium, 2=heavy);
date of final harvesting (eg., Nov 08, 2014).

The scales of these variables are:
(a) scale, **ordinal**, scale, scale
(b) scale, **nominal**, **nominal**, scale
(c) scale , **nominal**, **ordinal**, scale
(d) scale, **ordinal**, **ordinal**, scale

7. A study was conducted and the following variables were measured:
Smoking status (1=yes, 2=no),
Birth date (e.g., 11/07/2007).

The measurement of these variables are:
(a) scale, scale
(b) **nominal**, scale
 (c) scale, **ordinal**
(d) **ordinal, ordinal**
(e) scale, nominal

8. An experiment was performed on lab mice to investigate the effect of ingesting chemicals or pesticides sprayed on avocados upon subsequent cancer rates. The following variables were measured:

> gender (0=female, 1=male); weight (g); dose of chemical (none, low, high); number of tumors

Which of the following is FALSE?

(a) Gender is nominal; dose is ordinal
(b) Gender is discrete; weight is continuous
(c) Number of tumors is discrete
(d) Dose is ordinal, and gender is qualitative
(e) Weight is ratio; and number of tumors is discrete.

9. A random sample of 500 households in San Diego was selected and several variables are recorded for each household. Which of the following is NOT CORRECT?
(a) Household total income is a scale variable.
(b) Socioeconomic status was coded as 1=low income, 2=middle income, 3=high income. It is a scale variable.
(c) The primary language used at home (such as English, Spanish) is a nominal variable.
(e) The number of persons in the household (1, 2, 3, etc) is a discrete variable.

10. An experiment was performed on lab mice to investigate the effect of ingesting chemicals or pesticides sprayed on avocados upon subsequent cancer rates. The following variables were measured:

> gender (0=female, 1=male); weight (g); dose of chemical (none, low, high); number of tumors

Which of the following is NOT CORRECT?

(a) Gender is qualitative, Weight is quantitative, number of tumors is quantitative, dose is qualitative
(b) Weight is quantitative and continuous; number of tumors is quantitative and discrete
(c) Gender is qualitative and discrete, Weight is quantitative and continuous, number of tumors is quantitative and discrete, dose is qualitative and discrete
(d) Gender is nominal, categorical and qualitative
(e) Weight is quantitative, scale, ratio, and continuous

Chapter 2

Starting SPSS for Windows

Click on the start or Microsoft window button in the lower left corner of your computer, then on *Programs*, IBM SPSS Statistics, then IBM SPSS Statistics 22, depending on the version you have installed.   START>PROGRAMS>IBM SPSS Statistics>IBM SPSS Statistics 22.

When started (and it can take a while to show itself especially in the labs), SPSS displays the following:

Choose New Dataset if you need to type in your data or Open another file… if you wish to open a data set, and browse for the file.

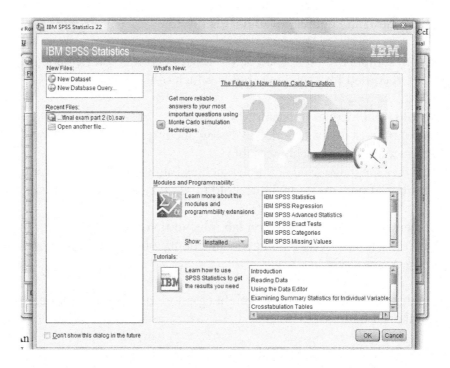

An alternative way of opening the most recently-used files is provided by **the Recent Files** options below the New Files option on the left.

SPSS Data Editor

The data editor window should now be displayed. Initially, there is no data in the data editor. We could begin typing data into the empty cells or open existing data.   We will begin to type in data soon.

First, let's look at some options and defaults (or make sure that they're already set), so that your work with SPSS runs smoothly.

Setting Some Options

To set the options, click on *Edit* at the top left of the screen, then on *Options* at the bottom of the resulting pull-down menu. The following display will appear.

1. An initial choice concerns the *Look and feel* of SPSS: Choose Standard so my screen shots match your program.
   a. Select Display names under Variable Lists
   b. Select Alphabetical under Variable Lists

Now click on the *Output Labels* tab.

You can choose whether to show *Names and Labels* (for variables) or *Values and Labels* (for *variable values*) appear on your output, meaning your graphs and charts. Traditionally, we choose only *Labels* for both boxes, but if you would like to show your value labels, this is where you make that adjustment.

Finally, after making all of your selections, choose APPLY and then OK.

Now, back to the data window, we can notice the Menu Bar at the top of the screen with the following options.

Menu Bar

- File – use this to open, read, print and save files as well as to exit SPSS.
- Edit – use this for editing functions: copy, paste, find, replace, etc.
- View – options that affect toolbars, status bars, fonts, and data or variable views.
- Data – relates to defining, configuring, and entering data; also, sorting cases, merging files, and selecting or weighting cases.
- Transform – use this to recode, compute new variables, reorder, categorize and deal with missing cases.
- Analyze – use for data analysis.
- Graphs – use for creating graphs or charts
- Utilities – mostly for advanced users, make complex data operations easier.
- Windows – deals with the positions, status, and format of open windows.
- Help – very useful aid with search capabilities, tutorials, and a statistics coach to help you with data analysis.

Opening a File –

Open SPSS first, then choose File>Open Data.

IMPORTANT - If you double click on an SPSS file in Windows Explorer, you may need to open SPSS first.

Data files
1. Select file, open, data or click the icon on the far left (file open).
2. SPSS automatically looks for files with the extension .sav. SPSS save (.sav) files contain data stored in compressed binary format.
3. Select the file and click open or double click on the highlighted file.

To open a data file click this icon

	age	marital	address	income	inccat	car	carcat	ed	employ	retire
1	55	1	12	72.00	3.00	36.20	3.00	1	23	0
2	56	0	29	153.00	4.00	76.90	3.00	1	35	0
3	28	1	9	28.00	2.00	13.70	1.00	3	4	0
4	24	1	4	26.00	2.00	12.50	1.00	4	0	0
5	25	0	2	23.00	1.00	11.30	1.00	2	5	0
6	45	1	9	76.00	4.00	37.20	3.00	3	13	0
7	42	0	19	40.00	2.00	19.80	2.00	3	10	0
8	35	0	15	57.00	3.00	28.20	2.00	2	1	0
9	46	0	26	24.00	1.00	12.20	1.00	1	11	0
10	34	1	0	89.00	4.00	46.10	3.00	3	12	0

Output files
1. Select file, open, output. (See picture below)
2. SPSS automatically looks for files with the extension .spo. These files contain the information produced by SPSS statistical procedures.

3. Select the file and click open or double click on the highlighted file.

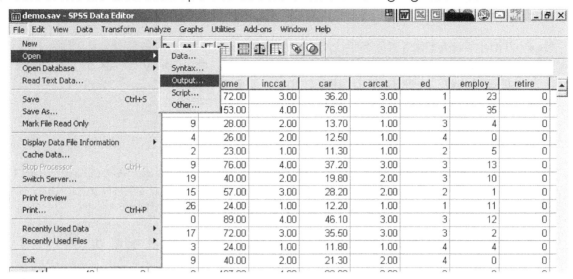

SPSS Tutorials (See picture below)

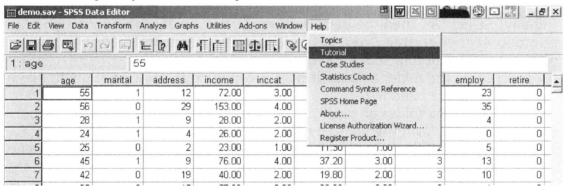

Go to the menu command Help> then choose Tutorial

Some good tutorials to begin with:
1. **Under Introduction, review opening a data file.**
2. **Under Using the Data Editor**
   a. **Entering Numeric Data**
   b. **Entering String Data**
   c. **Defining Data**
3. Finally, in the Tutorial on Working with Output, **Part 1 only, Using the Viewer.**

The Output Window
Output, or the result of previously conducted analysis, is the objective of all of our analysis work. Output can be very lengthy.

The outline view on the left of the screen is like a table of contents for the output. You can navigate the output by pointing and clicking on the element you wish to examine. To collapse a group of output objects, click on the minus sign to the left of the heading.

- Title – refers to the title of the procedure
- Notes – hidden in the output window, but may be viewed by double clicking on the closed book icon to the right of the notes title.
- Statistics, Frequencies, etc. – Output of interest.

Editing Output

You can edit output also. To move a portion of the output, select the object, then select edit, cut. Then select another output object below which you want to place the output object you have cut, select edit, then paste after.

To delete a certain section, select the section, then select edit, delete.

The insert text icon is particularly useful to make notes to yourself about the output.

Printing Output

Initially, Select file, then the page setup option. Specify the orientation of your printout – either portrait (vertical) or landscape (horizontal). You must select the portions of your output that you want printed, as follows:
1. Click on the folder or output object you want to print on the left side of the spss output window;
2. Click on the folder or output object on the left side of the spss output window, then hold down the shift key while selecting another folder or object – selecting all the both the objects you clicked and all the objects in between;
3. Or click on a folder or object on the left side of the output window, then hold down the control key while selecting additional objects - this selects only the objects you click on.
4. Select File, Print. Under Print Range, change radio button selection to All visible output – if you want to print everything. But BE CAREFUL, you may end up with pages and pages of output!

5. Then click file, print, and ok.

Exiting SPSS
Choose file from the menu bar and then exit. Or click the close button in the very upper right corner of the screen (the button with the x on it). If you haven't saved your work, you will be prompted to save the contents of each window (data, output, and perhaps syntax) that are currently open.

Limitations of the Student Version
The following limitations apply to the IBM® SPSS® Statistics 22 Student Version:
☐ Data files cannot contain more than 50 variables.
☐ Data files cannot contain more than 1,500 cases.

Entering Data into SPSS

Goals:

1. To be able to create an SPSS data file using the data editor to input data.
2. To learn to customize your data with variable labels and value labels so you can have a feel for what the question was about without always referring to the questionnaire.
3. In class – we will complete 2 data entry assignments - Data entry 1 and Data entry 2 and save to your flash drive.

Basic Steps in Creating a Data File

1) Give the file a name.
   i)   The rules for this are the same as for any Windows program, but the extension (the file name after the period) is always .SAV for an SPSS for Windows data file.
2) Each question or item needs a name, called a "variable name."
   a) Variable name rules:
      i) Variable name can be no longer than 64 characters, but I recommend you choose a much shorter (8 letters or fewer) variable name.
      ii) Variable names cannot end with a period.
      iii) Variable names must begin with a letter. The remaining characters can be any letter, any digit, a period, or the symbols @, #, _, or $.
      iv) Avoid variable names that end with an underscore.
      v) Variable names cannot contain blanks or special characters (!,?,' and *).
      vi) Variable names are not case sensitive.
      vii)   Each variable name must be unique.

Student sample quiz questions:

Which of the following are valid variable names in SPSS ? Which are not? Why?

---

i) FIRSTNAME
ii) ID   id
iii) U.S.Law   US Law   us_law
iv) State's
v) $price   price$

14

Solution to 2b questions:

i)	Valid.
ii)	Each is okay on their own. But they could not be used in the same data set. Remember, SPSS variable names are not case sensitive and must be unique.
iii)	U.S.Law and us_law are valid; US Law is not valid. Blanks are not allowed.
iv)	Not valid. See rule v) regarding '.
v)	$price is not valid. See rule iii). Price$ is valid.

3) Enter the data.
   a) If the data deal with the responses to a question in a questionnaire, the data will be codes that represent the answers.
   b) Use numbers only.
   c) If respondents don't answer a question, give more than one answer, or do something else that makes their answers unusable, we should code their response as "missing". We often use –1 or 99 as the code for "missing data". This is assuming that –1 or 99 is not a real, valid response to the question.
      i) For instance, if you asked someone how many pounds they lost this year, they might answer 99 – so 99 would not be a good choice for a missing value for that variable.
4) Save the data file.

There are other, optional things that can be done with the SPSS data file:

1) Create an extended label for each variable.
2) Create a label for each value or response.
3) SPSS is capable of looking at a variety of types of information. For example, it can look at whole numbers, numbers with decimal points, numbers given in a currency format, etc. We can tell SPSS which of these to use, how large the number is and, if a decimal point is used, how many digits there are on the right side of the decimal.
4) Do we need to exclude certain values?

- Record or observation or case- simply a line in the data set.
- Field or variable- a column containing the data for a specific question or characteristic of each case.

Complete the SPSS Tutorial on Using the Data editor

Go to the menu command Help> then choose Tutorial
1. Click the + button on the left side of Using the Data Editor and **complete all lessons within Using the Data Editor**

Illustration and data entry practice

For this discussion, let's assume that we have run an experiment to test the hypothesis that writing an essay in favor of doubling tuition at SDSU would make people more accepting of an increase in tuition. The participants were randomly divided into two groups, one group wrote an essay about why tuition should be doubled and the other group wrote an essay about the value of intercollegiate sports. We then administered a questionnaire to gather some demographic information about the participants. The questionnaire also measured attitudes towards several issues, including the issue about doubling tuition. These variables are summarized in Table 1. The responses from 4 participants are shown in Table 2.

Table 1. Codebook for the Tuition Study

Name	Variable Type	Variable Label/ Value Labels
ID	Numeric 3.0	
Demographic data		
BIRTHDAY	Date (mm/dd/yyyy)	Date of birth of the respondent(mm/dd/yyyy)
GENDER	String 1	Gender of respondent/ "F" "FEMALE" "M" "MALE"
CLASS	Numeric 1.0	Year in college/ 1 'Freshman' 2 'Sophomore' 3 'Junior' 4 'Senior'
INCOME	Dollar 11.2	Monthly income of respondent in dollars and cents
Treatment Conditions (Independent Variables)		
ESSAY	Numeric 1.0	Wrote essay about / 1 "doubling tuition" 2 "intercollegiate athletics"
Dependent Variables		

SHOCK	Numeric 1.0	Tuition should not increase by more than 5% each yr/ 1 "Strongly Disagree" 2 "Disagree" 3 "Slightly Disagree" 4 "Slightly Agree" 5 "Agree" 6 "Strongly Agree" 9 "No Opinion on This Issue"
TUITION	Numeric 1.0	Tuition at SDSU should be doubled/ -same value labels as SHOCK
PHDSOC	Numeric 1.0	SDSU should have a Ph.D. program in Sociology/ -same value labels as SHOCK

The numbers in the Variable Type column indicate: whether the variable is a numeric, date, dollar, or string variable; the width of the variable; and for numeric variables, how many decimal places. The first number after numeric, date, dollar, and string variables is the maximum width of the variable (including the number of whole digits, the decimal point, if any, and the number of decimal digits, if any. The number after the decimal point is the number of decimal places for the variable. Note that string variables and date variables do not have decimal places.

For example, Numeric 5.3 would work for a number like 1.837

There are 4 spaces taken up by the numbers 1 8 3 7 and 1 place taken by the decimal point, for a total of 5 spaces. Three spaces are allowed for the decimal portion of the number 837.

1 . 8 3 7

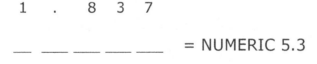

= NUMERIC 5.3

1   2   3   4   5   total spaces for the number including decimal point
NUMERIC 5 (the first part)

        1   2   3   spaces for the decimal portion of the number   .3 (the 2nd part)

Another example,   NUMERIC 8.2

Allows 8 spaces for all the number and decimal point, and 2 columns for the decimal number.

It would work for numbers like 243.42 (note only 6 spaces are being used here, 2 for decimal, 3 for the whole portion of the number and 1 for the decimal point)

It would also work for a number like 10243.88 (8 spaces are being used here)

Table 2. Data from 4 Participants in the Tuition Study

ID	BIRTHDAY	GENDER	CLASS	INCOME	ESSAY	SHOCK	TUITION	PHDSOC
1	3/20/1975	F	1	1823.62	1	5	2	6
2	5/32/1977	M	3	128.50	2	6	2	4
3	10/3/68	F	4	1239252	2	5	9	5
4	10/10/1582		2	879	1	9	9	5

There are no hard and fast rules about how to organize the variables within the data file. I typically begin with an ID number followed by demographic data (age, gender, etc.), the independent variable condition(s), and finally the responses to the dependent variables. The data in Table 2 follow that organization. You will normally be entering your data directly from the response sheets that you give to your participants. I don't recommend copying over the responses into a tabular format as in Table 2. The extra step of copying over the data onto a sheet to give to a data entry person who then enters the data into the computer will most surely cause errors to creep into the data.

Creating the data file for this example:

1) Give each case, e.g. each questionnaire, an identification number.
    a) Assists with checking the accuracy of our data entering.
    b) Creates a unique identifier for each questionnaire or person.
2) Create a variable name for each question. The variable name should be simple, but we want it to express the main idea of the variable in some way.   Refer to Table 2.
3) Add variable labels.
    a) The labels can be up to 256 characters long.
    b) For CLASS we could add a variable label of " Year in college".
4) Add value labels.
    a) Assign descriptive value labels for each value of a variable.
    b) Value labels can be up to 60 characters long.

Hands-on Training

To start creating the new data file, go into SPSS. The first thing you see is the Data Editor. It is set up like a spreadsheet, with the upper-left cell outlined.

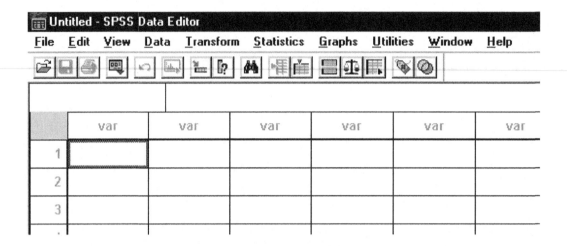

- Rows are for the cases, e.g., the respondents or the questionnaires
- Columns are for the variables, e.g., the questions
- Cells contain values. Each cell contains a single value of a variable for a case.
- The upper-left cell will usually contain the identification number for the first case and the cells across that row will contain data about that case.

Numeric Data

Entering simple numeric data is easy...Select a cell and enter the number. The data value appears in the cell and in the cell editor. Press Enter to record the value. If you haven't named the variable, The Data Editor assigns a unique variable name.

Let's begin entering our data!

ID (identification #) – our first variable name (and variable label)

a. Variable Name

Enter the variable definition information for each of the variables in Table 1. First, click on the variable view tab in the bottom left hand corner of your screen. In the 1st column (the upper left corner) of the worksheet in the Variable view window type **ID** as the variable name.   (See picture below.)

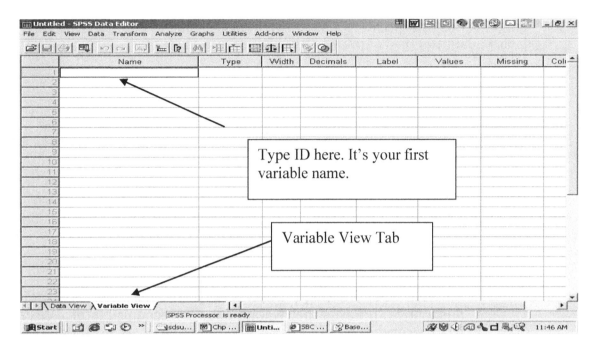

b. Variable Type

To change the variable type:

- Click in the appropriate cell in the Type column.

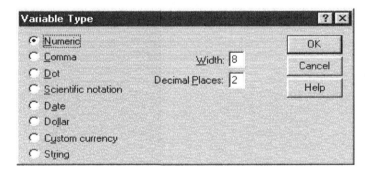

Click the small gray box (...) that appears in the cell to display the Variable Type box:

Specify the type by clicking the circle to the left of the appropriate type, Numeric.

Click OK after specifying the type. Change the decimal places to 0, since we have ID numbers that don't have
decimal points.

ID is a numeric variable.

More about the Variable TYPE column options

String is for any variable that contains 'alpha' characters, (anything other than numbers) or text as data. String variables can contain any letters, numbers, and symbols.

Numeric is the most common and is the default type.

Date variables typically consist of a year, month, and day. Typical date variables are date of birth, date and time of testing, etc.

For numeric variables, you can set the width (number of spaces reserved for variable when its values are displayed) and the number of decimal places in this dialogue box. These may also be set using the Width and Decimals columns on the Variable View sheet where clicking on a cell produces either a gray box or up/down arrows to make selections. (See picture below) The width does not refer to how many digits are stored in the data file, width refers to how many digits will be displayed in the data editor and in the output.

Width	Decimals
8	0 ⬍

c. Variable Label

Type in **Identification #** as the variable label.

Variable label is a longer description of the variable. Use brief, but descriptive, phrases that will be easy to recognize later. Variable labels will preserve the case (upper and lower case) as entered.

d. Value Labels

Value labels identify the coding scheme for the values. *Value labels* are not mandatory and they would not be used for ID values or other interval type data such as temperature values or scores on tests (e.g., you wouldn't label each value of an IQ score). *Value labels* are typically used when the value refers to a specific category such as "male" and "female," or the scale values for a response scale, e.g., "strongly agree." Let's leave the *Labels* section blank for the ID variable.

e. Missing Values

Because you assign the values of ID there are "no missing values" or questionnaires or subjects without an ID number.

g. Measurement

Measurement refers to scale of measurement: nominal, ordinal, interval, or ratio. SPSS allows you to assign one of three categories of measurement:

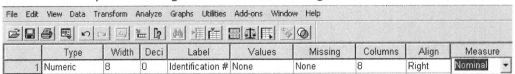

	Type	Width	Deci	Label	Values	Missing	Columns	Align	Measure
1	Numeric	8	0	Identification #	None	None	8	Right	Nominal ▾

nominal, ordinal, or scale. ID is a nominal variable. Your screen should look like the picture here.

Now toggle or click the tab on the bottom right of your screen for the **data view** window and enter the ID data, given in Table 2.

Table 2. Data from 4 Participants in the Tuition Study

ID	BIRTHDAY	GENDER	CLASS	INCOME	ESSAY	SHOCK	TUITION	PHDSOC
1	3/20/1975	F	1	1823.62	1	5	2	6
2	5/32/1977	M	3	128.50	2	6	2	4
3	10/3/68	F	4	1239252	2	5	9	5
4	10/10/1582		2	879	1	9	9	5

Our second variable is *BIRTHDAY*

a. Variable Name- Move to the second row in the variable view window. Enter BIRTHDAY as the variable name.

b. Variable Type- Date. Select *mm/dd/yyyy* from the date options that appear and click the ok button.

c. Variable Label – Date of Birth

d. Value Label-Not needed.

e. Missing Values

The easiest way to deal with missing values for date variables is to just leave the value blank. A blank numeric value will be displayed as a single period(.). Try deleting one of the date values and note that the result is a period for that case. A blank numeric value or a blank date value is defined as a **system missing value**. User-missing values are not available for date formats!

f. Columns – Change your column size to 10. This allows us to see the full date in our data view window. Notice the Date type takes 10 column spaces.

g. Measurement - The level of measurement for a date variable is "scale."

	Name	Type	Width	Deci	Label	Values	Missing	Columns	Align	Measure
1	ID	Numeric	8	0	Identification #	None	None	8	Right	Nominal
2	Birthday	Date	10	0	Date of Birth	None	None	8	Right	Scale

Change this to 10.

Enter the data.

Switch to data view window and enter the first birthday value from Table 2 (3/20/1975).

Enter the date for case #2 (5/32/1977). The Data Editor should show a ., refusing to enter a value for date that is not possible. Reenter the date as 5/30/1977.

Enter the date for case #3 (10/3/68). Note that the year in the data file has been changed to 1968, while the display remains at 68. If you enter a 2-digit year, SPSS will automatically add "19" to the year.

Enter the date for case #4 (10/10/1582). The Data Editor does not enter the date. Why? SPSS stores the date as the number of seconds from October 14, 1582 (the beginning of the Julian calendar). As a consequence you cannot enter a date that is on or before October 14, 1582. Try to enter the date as 10/25/1582. Your data should appear as the picture here.

	ID	Birthday
1	1	03/20/1975
2	2	05/30/1977
3	3	10/03/1968
4	4	10/25/1582

GENDER – Normally I would code gender as a numeric variable (i.e. 1=male, 2=female), but for illustration purposes we are coding gender as a string variable.

Caution on string variables: You are very limited on the types of analysis you can do with string variables. I suggest coding variables as numeric whenever possible.

a. Variable Name

Click on the 3rd variable row in the SPSS Variable view window. Enter GENDER as the name for the third variable.

b. Variable Type

Click *string* as the variable type. Because only one letter is needed to enter the M and F codes, enter 1 as the number of characters. The term "string" variable is synonymous with "alphanumeric" variable. The data editor will only allow you to enter as many characters as you have defined in this dialog box. Because the number of characters was defined as 1, you will only be allowed to enter single characters.

c. Variable Label - Enter a variable label such as "Gender of the Respondent."

d. Value Label

SPSS is sensitive to the case of string values. The value "F" is different from the value "f". Suppose you decide to enter the value M for males and the value F for females in the data file. Then you can attach the value label "male" to the letter M and the value label "female" to the letter F.

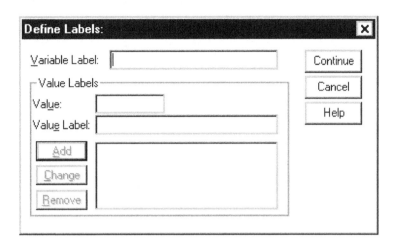

After naming and labeling the variable, give each possible response a value name and label the values in a way that would be useful. For example, using the variable GENDER again, M would be for male, so type M in the value box, and "MALE" in the value label box, and then click "Add". Repeat for the female value codes. (When you want to modify value labels, click " **Change**," and click "**Remove**" if you want to delete one.)

e. Missing Values

You will notice in Table 2 that the 4th person did not give their gender. Therefore, we have a missing value.

Table 2. Data from 4 Participants in the Tuition Study

ID	BIRTHDAY	GENDER	CLASS	INCOME	ESSAY	SHOCK	TUITION	PHDSOC
1	3/20/1975	F	1	1823.62	1	5	2	6
2	5/32/1977	M	3	128.50	2	6	2	4
3	10/3/68	F	4	1239252	2	5	9	5
4	10/10/1582		2	879	1	9	9	5

There is no such thing as a **system missing value** for a string variable. Blank string values are considered to be valid values. If you have missing data for gender and happen to leave the value blank, SPSS will think you have three valid genders, M, F, and blank. We will create a **user-defined missing value** for the variable Gender. If the person did not answer the question, we give them a value of 9 for missing or no answer.

To enter user-defined missing values click in the variable view window, place your cursor in the column-

Missing Values... then choose
    Discrete missing
values

Your cursor should be in the first of the three boxes, input 9, and then click the Ok button. You have now defined a 9 as a user missing value.

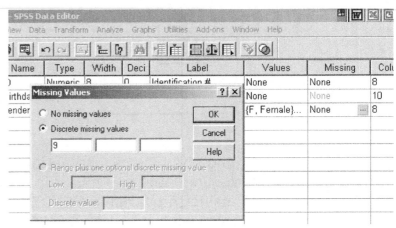

We can add a value to our 9 in the value labels section. Click on the 3 dots in the Values Column.   1. Type in the value of 9 for our missing value indicator

2. the value label of Missing,

3. Click Add and

4. Ok.

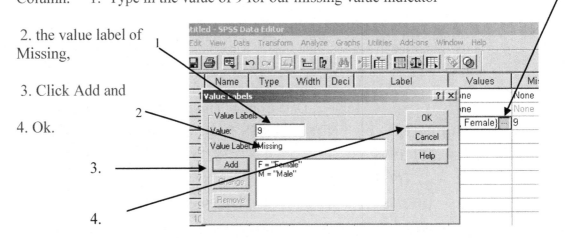

Note: SPSS assumes that all string variables are left justified. Numeric variables are right justified.

g. Measurement

When you identified the variable type as "string," the measurement type was automatically set to "nominal." Nominal is the correct measurement type for gender.

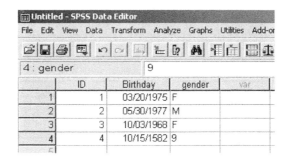

Enter the data.

Now, in the Data View window, Go to the menu command **View**> and choose **Value Labels**. Your data should now look like the picture shown. We can now see the value labels associated with Gender, rather than M,F, and 9.

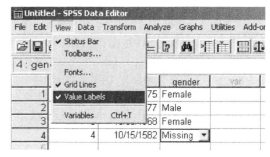

CLASS (integer numeric with value labels)

a. Variable Name -Click on the 4th variable row in the variable view window. Enter CLASS as the variable name.

b. Variable Type- Numeric1.0 (width = 1, decimal places = 0)

c. Variable Label-"Year in school?"

d. Value Label- Enter the four value labels: 1 = 'Freshman', 2 = 'Sophomore', 3 = 'Junior', and 4 = 'Senior.'

*e. Missing Value-*Missing values should be left blank. We do not have any missing values in Table 2.

g. Measurement

What scale of measurement is "year in school?" It is at least ordinal, a person has earned more credit hours if he or she is a senior than if he or she is a junior. This is an ordinal variable.

Continue to enter the remaining variables on your own.

Once the variable names and labels and the value names and labels are Give your new data a file name and save it before you exit or go on to using the file. On the first screen, you can save by clicking on "File" and using "Save As" to enter your file name. After you have named the file and saved the data the first time, you can save changes with "File" and " Save" or by clicking the little disk icon near the upper-left corner of the screen. It is important to save data frequently.

> You can insert a new case or a new variable. Move the cursor to the row or column where you want to make the insertion and then right click and then click either *Insert Case* or *Insert Variable.*

Check the accuracy of your data by skimming down each column for codes that are impossible with these value labels. For example, Gender can have only three possibilities since males are M, females are F, and missing information is 9. You could do this on the screen or on a frequency distribution from SPSS.

IMPORTANT: DO NOT EDIT AN SPSS FILE WITH A WORD OR TEXT PROCESSOR.

An SPSS system file is a specially formatted file that should not be edited with a word or text processor. If you try to edit a systems file outside of the SPSS Data Editor you may destroy structure of the file and you may not be able to open it again within SPSS. If you happen to open an SPSS systems file with a word processor close the file without making any changes to it.

Chapter 3

Frequency Charts and Descriptive Statistics

<u>Goals:</u>
1. **Create frequency distributions in SPSS.**
2. **Create bar charts and histograms using the Frequency procedure.**
3. **Produce descriptive statistics including, mean, median, mode, variance, standard deviation, skewness, kurtosis, maximum, minimum, range, sum and standard error.**

The Organization of Information – Frequency Distributions
- *Social researchers – gather larger amount of data from surveys, maybe thousands of individual responses i.e. GSS data*
- *How do we summarize this to make any conclusions?*

Frequency chart or table– a table reporting the number of observations falling into each category of the variable.

- *This is usually the first step in data analysis.*
- *After we enter our data, we run frequency charts to make sure the categories make sense, and every value falls into the appropriate category.*

There are several columns in a frequency table, including Count or Frequency, Proportion, Percentage, Cumulative Frequency, and Cumulative Percentage.

Frequency (f) = the count of cases or observations in that particular category
Proportion (P) – a relative frequency obtained by dividing the frequency in each category by the total sample size (N).   The formula is P=f/N
*Percentage (%) = Proportion * 100.   The formula is P * 100*
Cumulative Frequency (CF)= The sum of all frequencies above & including this row.
Cumulative Percentage (%)= The sum of all percentages above & including this row.

Notes: 1. The sum of all frequencies = N (total sample size)
 2. The sum of all proportions = 1.0
 3. The sum of all percentages = 100
 4. The last row of the cumulative frequency must = N and the last row of the cumulative percentage must equal 100.

Example: In 1993, 36 states and Washington, D.C. had statutes permitting capital punishment. Of these 36 states, 27 set a minimum age for execution. Assume you are a member of a legal reform group that is trying to get the states that do not have a minimum age for execution to change their laws. *You want to prepare a report describing the minimum age for execution in the 27 states.*

State	Minimum Age	State	Minimum Age
Arkansas	14	Texas	17
Virginia	15	California	18
Alabama	16	Colorado	18
Delaware	16	Connecticut	18
Indiana	16	Illinois	18
Kentucky	16	Louisiana	18
Mississippi	16	Maryland	18
Missouri	16	Nebraska	18
Nevada	16	New Jersey	18
Oklahoma	16	New Mexico	18
Wyoming	16	Ohio	18
Georgia	17	Oregon	18
New Hampshire	17	Tennessee	18
North Carolina	17		

Steps for Frequency Table Creation
1. *List categories - In our case 14, 15, 16, 17, 18*
2. *Tally the number in each*
3. *Write down the frequency (add up your tallies)*
4. *Calculate the proportion*
5. *Calculate the percentage*
6. *Last row – include your total N (sample size) and make sure your proportions sum to 1 and your percentages sum to 100% (with a little adjustment for rounding if necessary)*
7. Create the cumulative percentage and cumulative frequency columns.

Category (X)	Frequency	Proportion	Percentage	Cumulative Freq	Cumulative %
14					
15					
16					
17					
18					
TOTAL					

Now that you have had a chance to explore SPSS and have learned to enter and define data, you can start exploring statistical procedures. The calculation of basic descriptive statistics is a good place to begin, when you start your analysis. Summarizing and describing your data using frequency tables, measures of central tendency (mean, median, mode), and variability will help you organize and report your results clearly.

We will begin by using frequencies and percentages to summarize your data. This is useful for determining how many people in your sample have a particular score or belong to a group within a variable. For example, we might want to know what percentage of students are males and what percentage are females.

Another reason for using frequency charts on your data is to check the accuracy of your data entry. For example, if you coded gender with 1 or 2 for male/female, and you notice a 3 in your frequency chart, you have found a mistake in your data entry. You can go back to your data view window, find the errors and make the necessary corrections.

The DATA

The data contains information about graduate students. The variables are as follows:

EDUC	Degree completed	1=bachelors, 2=masters, 3=specialist, 4=doctorate
TEAEXP	Have you had previous teaching experience?	0=no,1=yes
DEGREE	Degree program enrolled in	1=bachelors, 2=masters, 3=specialist, 4=doctorate
GENDER	Gender of respondent	1=male, 2=female -1=missing
COMPEXP	Computer experience	
GROUP	Class section	Defined as follows, 672=class numerical code, f92 refers to fall of 92, and so forth
AGE	Age of respondent	

Questions we will answer

1. How many students are male and how many students are female?
2. What % of students are male and what % are female?
3. Which section had the largest enrollment?
4. How much computer experience do students have when they enroll in a beginning statistics class (672)?
5. What type of degree are these students pursuing?

Let's begin by opening our data

We will use the data found in blackboard called Frequencydata.sav. You can find the data set under Course Documents> Data Sets for SPSS Lab book. Download this file to your computer or save to your usb. Then, Start SPSS (See Chp 2 if you need help). From the main menu, select FILE>Open> Data and find the file Chp 4 data on your computer or USB.

Running Frequencies using SPSS

Step 1 Go to the Analyze menu command.
Step 2 Select Analyze, then choose Descriptive Statistics.
Step 3 Then select Frequencies.

These steps are shown in the pictures below.

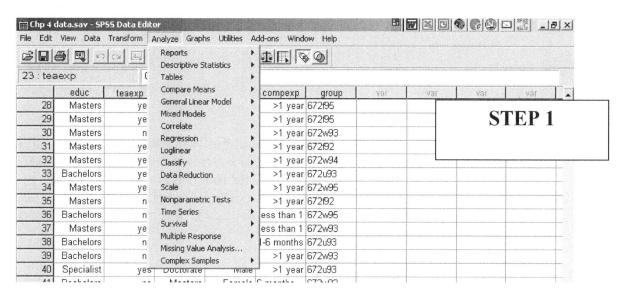

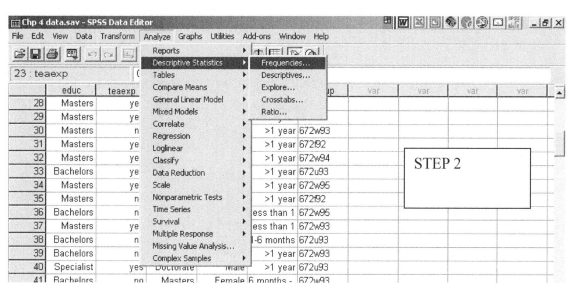

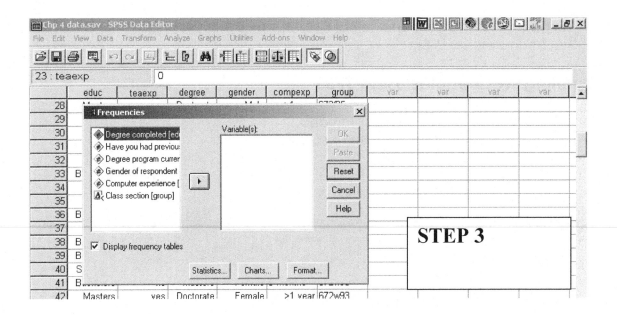

Once you have completed STEP 3, you must select the variables you are interested in. All possible variables are listed in the left side of the box, shown in the picture on the previous page. Next to this listing of all variables, there is a box labeled Variable(s). The variables placed in this box are your selected variables, or variables of interest. You may transfer the variable(s) of interest into the Variables box by highlighting the desired variable and clicking on the arrow button.

Remember, our first question involves the variable GENDER

1. How many students are male and how many students are female?

So, let's run a frequency chart of GENDER to determine the answer to question 1. Move

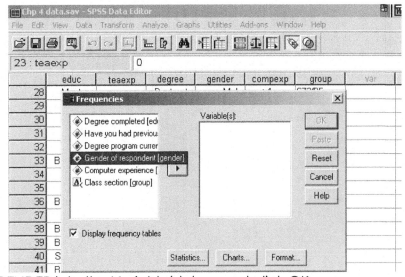

GENDER into the Variable(s): box and click OK.

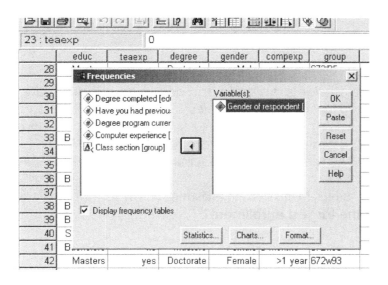

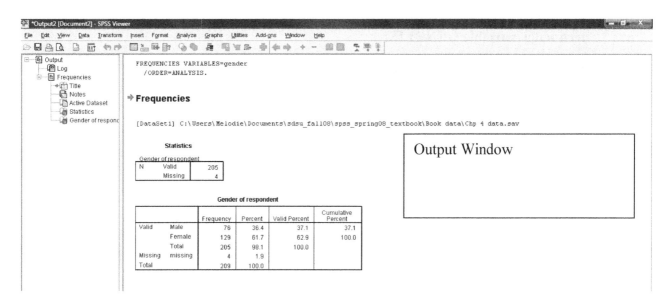

Notice in your output, that the first thing shown is a Statistics box. This box tells how many cases or observations or people reported their gender (VALID) and how many were missing or didn't report their gender. There are 205 valid responses and 4 missing responses for this data.

The next box shown is labeled Gender of Respondent (or GENDER depending on your option settings). The values of gender are listed in the left side of the table. In the next column, the Frequency of each gender is given. There are 76 males and 129 females in our sample data. The next column, labeled Percent, shows the percent represented by each group. For our data, 36.4% are male, 61.7% are female and 1.9% are missing or not given. The next column, Valid Percent, is a percentage based on the number of people who answered the question. If there are no missing cases, the Valid Percent and Percent columns will be the same. **FOR THIS CLASS, you will always report the VALID PERCENT!** Finally, the column labeled Cumulative Percent is calculated by adding up the valid percentages as you move down the list of categories. For the last category, the cumulative percentage is 100%.

So, using the above output, we can answer our first question:

1. How many students are male and how many students are female?
   There are 76 males and 129 females.

2. What % are male and what % are female?
   There are 37.1% male and 62.9% female. (REMEMER REPORT VALID PERCENT)

We can use frequency tables to answer the questions below as well. Let's complete question 3, **Which section had the largest enrollment?**

Here are the steps:

Analyze>Descriptive Statistics>Frequencies

Click the RESET button on the right side of your pop-up box, to remove the previous frequency variable GENDER.

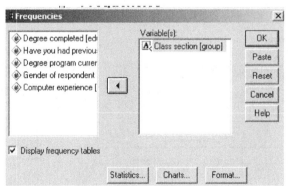

Move the variable GROUP or Class section, into the Variable(s): box, and click OK.

Class section		Frequency	Percent	Valid Percent	Cumulative Percent
Valid	672f92	35	16.7	16.7	16.7
	672f95	18	8.6	8.6	25.4
	672u93	18	8.6	8.6	34.0
	672u95	21	10.0	10.0	44.0
	672w92	26	12.4	12.4	56.5
	672w93	32	15.3	15.3	71.8
	672w94	21	10.0	10.0	81.8
	672w95	38	18.2	18.2	100.0
	Total	209	100.0	100.0	

Resulting Frequency Chart or Table

3. Which section had the largest enrollment? *Section 672w95 had 38 students, making it the largest section. You could also look under Valid Percent and see that*

35

section 672w95 had 18.2% of all students, the largest percentage compared to the other sections.

Complete the remaining questions on your own, using the frequency command. The answers are given below.

4. How much computer experience do students have when they enroll in a beginning statistics class (672)?
5. What type of degree are these students pursuing?

Solutions:

Question 3: 18.2% have fewer than 6 months experience. Then using the Cumulative %, we can say 69.4% had more than 1 year of computer experience.

Question 4: 33.5% are pursuing a masters degree, 3.8% are pursuing a specialist degree, and 62.7% are pursuing a doctorate degree.

Summary of Frequency charts:

SPSS command: **Analyze>Descriptive Statistics>Frequencies>choose variable >ok.**

Often it is helpful to begin your analysis of nominal or ordinal level variables by generating a frequency table.

A simple frequency distribution table displays the number and percentage of times each score appears in a data set. The total of all the individual frequencies equals the number of cases (subjects) in the data set.

- **Usually only the "Frequency" and "Valid Percent" columns are reported in the literature**
- **Values reported in the "Percent" and "Valid Percent" columns are identical unless there is missing data**
- *Often the row listing the number of missing cases is excluded. Instead the researcher will indicate the number of valid cases in the title and/or within the body of the paper.*

Descriptive Statistics

Descriptive statistics are used to describe or summarize the important characteristics of a data set.

Measures of Center (or Central Tendency)

A measure of central tendency indicates the location of the bulk of scores in a distribution. We are looking for the center or middle of the data set. The three main measures of central tendency are the mean, median and mode.

Mode

The mode is the **least informative of the measures of central tendency as it simply indicates the score that occurs most frequently. An advantage of the mode is that it can be reported with any level of measurement, nominal, ordinal or scale variables. Generally the mode is very unstable as it can change dramatically with relatively minor changes in the data set.**

In some data sets, more than one score may have a high frequency. In these situations, two modes may be identified (bimodal distribution) or multiple modes (multi-modal).

Median

Other names for the median are the 50th percentile or middle score. Fifty percent of the distribution is at or below the 50th percentile.

If you were to calculate this value by hand, you would arrange all the scores obtained for a variable from lowest to highest. The midpoint of this distribution is the score (value) at which half the observations are smaller and the other half are larger.

The median tends to be a better measure of central tendency than the mode because:

- *only one score can be the median*
- *median will usually be located close to where most of distribution is located*
- *The median is not used to describe nominal data, however it is an appropriate measure of central tendency for ordinal level data. It may also be a better measure of central tendency than the mean for interval or ratio-level data that have a skewed distribution.*
- *The distinguishing feature of the median is its stability when exposed to outliers (extreme scores). It is not impacted by outliers.*
- *A limitation associated with the use of the median is that it does not consider the mathematical values of the actual scores.*

Mean

The mean is the most frequently used measure of central tendency. The mean is also referred to as the arithmetic average as it is located at the mathematical center of a distribution. Conceptually, it can be viewed as the balance point for a distribution. It is the point at which the distribution is balanced (just like a teeter-totter).

One advantage of the mean is that it takes into account all of the available information when computing the central tendency of a frequency distribution. It is calculated by adding the scores or values for each subject, and then dividing this summative score by the total number of subjects.

The distance separating each score from the mean is called the deviation from the mean. The deviation score provide information about a score's relative location in a distribution. If the deviation score is positive than the score is to the right of the mean (greater than), if negative it is to the left of the mean (less than). A unique feature of the mean is that if you sum the deviations around the mean for all the scores the total will always equals zero. No other value in the distribution will do this. Most of the formulas for parametric statistics use the mean and the sum of the deviations around the mean.

The mean is an appropriate measure of central tendency for interval or ratio level variables. In addition, it may be used for ordinal level variables that have several values/categories.

A limitation associated with the use of the mean is that it is very sensitive to extreme scores.

Example

If you had a data set with five scores:

$$5, 12, 14, 22, 7$$

and you calculated the mean for this data set (add up all the values and divide by number of scores), you should end up with a mean of 12.

But look what happens to the mean if you change the last number from a "7" to "67":

$$5, 12, 14, 22, 67$$

The mean is now equal to 24.

The mean has doubled in value even though only one score in the data set has changed.

Note: How the mean score moves in the direction of the extreme score. This will always be the case.

Generating a Mean, Median, and Mode using SPSS

We will again use the same steps we used to generate a frequency table, as follows, adding 3 additional steps to generate the statistics:

Step 1 Go to the Analyze menu command.
Step 2 Select Analyze, then choose Descriptive Statistics.
Step 3 Then select Frequencies.
Step 4 Move the Variable AGE or Age
Step 5 Press the statistics button at the
Mean, Median, and Mode.
Step 6 Press Continue, and then Press

SPSS results. The mean age of our respondents is 36.48. The median age is 36, and the mode is 43. There were 209 people in our data set.

| Mode | 43 |

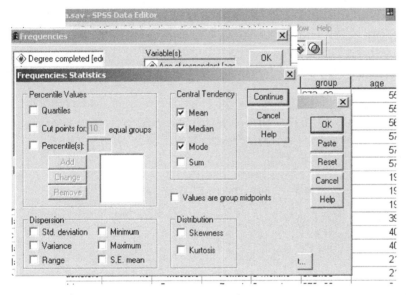

Measures of Variability

Variability refers to the spread or dispersion of the scores. If all scores in a distribution are the same, there is no variability. If all the scores are different and widely spaced, the variability will be high.

Minimum-Maximum Score

Perhaps the simplest way to describe the variability in a distribution is to report the minimum and maximum scores. A limitation of this approach is that one extreme score can significantly alter the results.

Range

Occasionally, you will see the range of a distribution reported. Subtracting the smallest score from the highest score creates the range.

Deviation Scores

In the discussion of the mean, we discussed that if you subtracted each value in a distribution from the mean and then added each of these scores, the total will always equal zero (Sum of the Deviation Scores). Consequently, this summative score is of no value in describing the variability in a distribution. However statisticians have found a way to utilize these deviation scores simply by squaring the values. These squared values are used to compute the Variance and Standard Deviation of the distribution.

Variance & Standard Deviation

When the mean is reported as the measure of central tendency, the variance and standard deviations are the appropriate measures of variability. These measures are comparable to the mean as they consider each score in the distribution. They are calculated by measuring how much the scores differ from the mean.

Variance

Variance is calculated by squaring and then summing all the deviation scores, and then dividing by the total number of cases (N). In other words, the variance of a set of observations is the average of the squares of the deviations of the scores from the mean.

Squaring all the deviation scores effectively eliminates all negative numbers from the equation. So if you ever obtain a negative variance, you will know that you've made an error.

A limitation of the variance is that the squared deviation scores no longer reflect the original scale of measurement. Consequently, the value is difficult to interpret. For example, if we find the customer waiting times at Washington Mutual bank in minutes, the variance will have the units in square minutes (min^2). This is not very useful or easy to understand. Once again, statisticians have identified a strategy to rectify this problem - it is called the "standard deviation".

Standard Deviation

Once the variance has been calculated, a simple mathematical procedure is used to return the deviation score to the original scale. Calculating the square root of the variance forms the standard deviation.

The standard deviation indicates how much the scores deviate above and below the mean. The larger the value, the more the scores are spread out around the mean, and the wider the distribution.

The standard deviation is the most commonly reported measure of variability at least when the distribution of scores is normal.

Generating a Variance, Standard Deviation, Minimum, Maximum, and Range using SPSS

We will again use the same steps we used to generate a frequency table, as follows, adding 3 additional steps to generate the statistics:

Step 1 Go to the Analyze menu command.
Step 2 Select Analyze, then choose Descriptive Statistics.
Step 3 Then select Frequencies.
Step 4 Move the Variable AGE or Age of Respondent into the Variables: box.
Step 5 Press the statistics button at the bottom of the pop up box and choose Variance, Standard Deviation, Minimum, Maximum, and Range.
Step 6 Press Continue, and then Press Ok.

Statistics

Age of respondent

N	Valid	209
	Missing	0
Std. Deviation		9.999
Variance		99.982
Range		39
Minimum		18
Maximum		57

SPSS Results. The standard deviation of the variable age is 9.999 years. The variance is 99.982. The range of ages was 39, with the minimum age of 18 and the maximum age of 57.

Shape of Distribution

Many statistical tests are based on the assumption that the data are normally distributed.

Normal Distribution

The normal distribution is the perhaps the most important distribution in statistics. All variables that are described as having a normal distribution have a shape that is a continuous, bell-shaped distribution. The normal curve is characterized by five properties

> **Characteristics of Normal Distribution**
>
> 1. **Unimodal (one hump in the middle)**
> 2. **Mean, median and mode are equal**
> 3. **Symmetrical. If fold curve in half, two sides fit perfectly**
> 4. **Asymptotic. Extremes never touch x axis (scores go to infinity)**
> 5. **Neither too peaked nor too flat (zero kurtosis)**

Skewness

Whenever data becomes skewed the mean is affected more then the median. The bottom graph shows how the mean and median are about the same on a normal distribution.

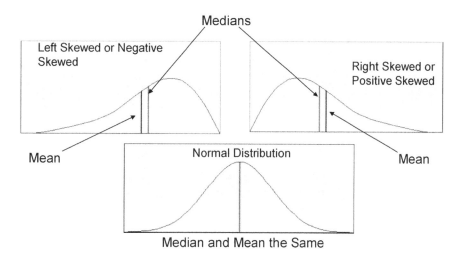

Skewness

Skewness refers to a lack of symmetry or balance in a frequency distribution. When a distribution is negatively skewed, the mean is smaller than the median as the mean always is pulled towards the outliers. The tail of the distribution will be stretched to the left. A positive skewness occurs when the mean is larger than the median. The tail of the distribution is stretched to the right (towards the positive numbers).

Kurtosis

Kurtosis indicates whether the peak of the distribution is taller or flatter than the ideal normal curve, and also whether the tails are shorter or longer than the normal curve. A positive kurtosis occurs when the distribution is very peaked. A negative kurtosis occurs when the distribution is flat.

Example of a Positive Kurtosis

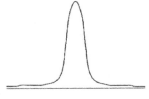

Peaked distribution

Example of a Negative Kurtosis

Flat distribution

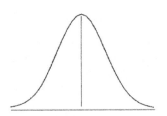

Normal Distribution   (Skewness = 0; Kurtosis = 0)

Generating Skewness and Kurtosis using SPSS

Step 1 Go to the Analyze menu command.
Step 2 Select Analyze, then choose Descriptive Statistics.
Step 3 Then select Frequencies.
Step 4 Move the Variable AGE or Age of Respondent into the Variables: box.
Step 5 Press the statistics button at the bottom of the pop up box and choose Skewness and Kurtosis.
Step 6 Press Continue, and then Press Ok.

Statistics

Age of respondent

N	Valid	209
	Missing	0
Skewness		.175
Std. Error of Skewness		.168
Kurtosis		-.848
Std. Error of Kurtosis		.335

SPSS Results. The distribution of age is about normal skewness (since .175 is pretty close to 0, we say it is approximately normal. **If the # is between -.5 and .5, we'll say that is close to 0 or about normal for skewness and kurtosis.** The data is flat with a kurtosis of -.848 (since the kurtosis is less than - 0.5, we say it is negative or flat).

The following table provides some general guidelines for selecting an appropriate numerical descriptive statistics.

Appropriate Descriptive Statistics for Variables with Different Levels of Measurement			
	Nominal	**Ordinal***	**Interval / Ratio**
Central Tendency			
1. Mean	No	Yes (Mean Rank)	Yes
2. Median	No	Yes	Yes
3. Mode	Yes	Yes	Yes
Variability			
1. Standard Deviation	No	Yes (of Ranks)	Yes
2. Minimum-Maximum (Range)	No	Yes	Yes
Shape			
1. Skewness	No	No	Yes
2. Kurtosis	No	No	Yes
Note. * If a researcher decides that an ordinal level variable "*approximates interval-level*", then she/he may use the descriptive statistics outlined in last column.			

Standard Error

SPSS computes the standard errors for the mean, the kurtosis, and the skewness. Standard error is designed to be a measure of stability or of sampling error.

Logic: You take a random sample from a population, you compute the mean. If you take another sample of the same size from the same population and again compute the mean. You get a slightly different number. If you collect many such samples, the standard error of the mean is the standard deviation of this sampling distribution of means. The logic behind computation of standard error for kurtosis and skewness is similar. A small value indicates greater stability or smaller sampling error.

Graphic Representation

Graphs can be extremely effective tools for enhancing understanding of the data. They are also an easy way to communicate the overall distribution of a set of scores.

When constructing a graphic representation of a variable, it is essential that you include sufficient information so that the graph (figure or table) is self-explanatory.

Essential information for all graphs includes:

- *A title*
- *Labels on each axis.*

The horizontal axis is referred to as the "X" axis (abscissa), and the vertical axis is referred to as the "Y" axis (ordinate).

Bar Graph & Pie Chart

Nominal *level variables are usually depicted using either a* <u>*Bar Graph or Pie Chart.*</u> **ORDINAL** *level data may be illustrated using these graphs if the variable has only a few categories(usually less than 10). If an ordinal variable has more than 10 categories, use a histogram.*

Using a Bar Graph, you can quickly compare the relative distribution of the various categories. The height of the bars shows the counts (or the percentage of cases) in each category. A space is inserted between each bar on the graph to emphasize that <u>*each category is distinct (non-overlapping)*</u>.

Let's generate a bar graph of GENDER.

STEPS TO GENERATE A BAR CHART USING THE FREQUENCY PROCEDURE IN SPSS
Step 1: Go to Analyze>Descriptive Statistics>Frequencies
Step 2: Select your variable and put it in the variables box (we will select gender)
Step 3: Push Charts button and select bar chart
Step 4: Choose either Frequencies (under chart values) or Percentages (we will use frequencies for our example)
Step 5: Press continue
Step 6: Unselect Display frequency table and Press OK. (unselecting display frequency table will allow us to see just the graph in our output, not the frequency table.

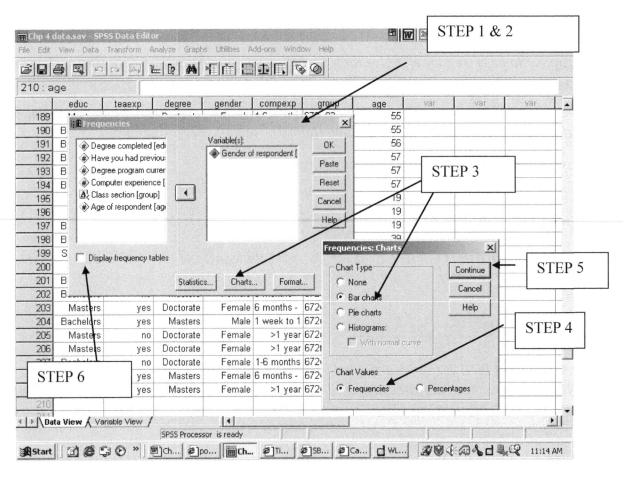

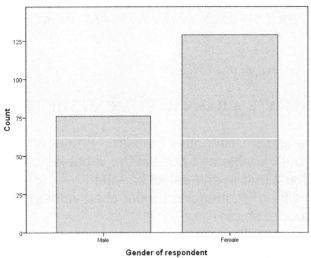

Gender of respondent

We can add labels to our bars by Double clicking on the chart in SPSS. This launches the Chart editor window.

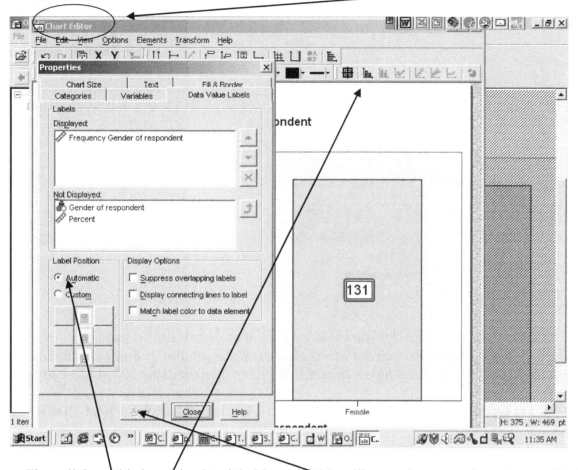

Then click on this icon, the data label button. This will open the pop up box shown, titled Properties.

Choose the Automatic Label Position button, Click Apply and Close. Then close the Chart editor window. The resulting graph is shown below.

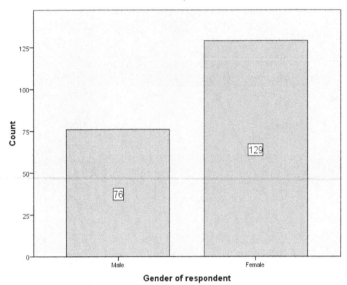

SPSS Results: There are 76 males and 129 females in our data set. The majority of our respondents are female.

The Pie Chart is useful as it helps you see what part of the whole each category represents. Construction of a pie chart requires the inclusion of all the categories that make up the whole. A pie chart should not be used unless the total for all responses equals 100 percent (a full pie).

STEPS TO GENERATE A PIE CHART USING THE FREQUENCY PROCEDURE IN SPSS

Step 1: Go to Analyze>Descriptive Statistics>Frequency

Step 2: Select your variable and put it in the variables box (we will select gender)

Step 3: Push Charts button and select pie chart

Step 4: Choose either Frequencies (under chart values) or Percentages (we will use percentages for our example)

Step 5: Press continue

Step 6: Unselect Display frequency table and Press OK. (unselecting display frequency table will allow us to see just the graph in our output, not the frequency table)

Step 7: To show the labels, double click on the pie chart in your SPSS output window. You should now be in the Chart Editor window.

Step 8: Choose the menu command Elements>Show Data labels. Click ok on the pop up box that appears and close your chart editor window.

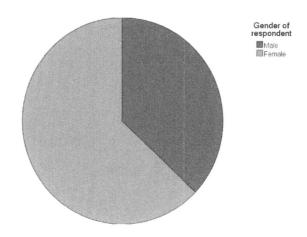

Gender of
respondent
■ Male
□ Female

Step 7 & 8 – Showing Data Labels on Pie Charts in SPSS

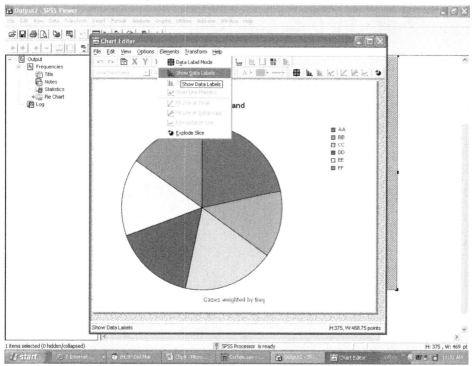

Once you have chosen to Show Data Labels, the Properties Box appears, and you can edit the size of your data labels on your pie chart. To edit the font size of your data labels, you choose the text style tab, and change your font size from 7 to 12 or more. This makes it easier to read your data labels.

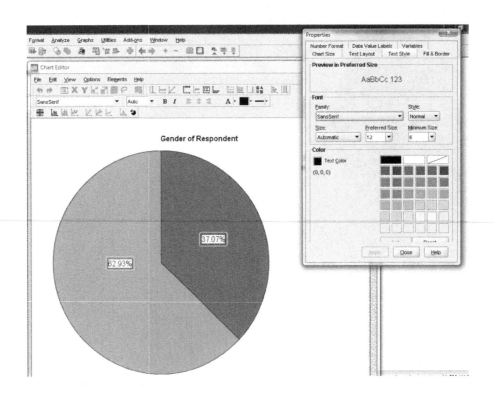

Edited Pie Chart with Data Labels

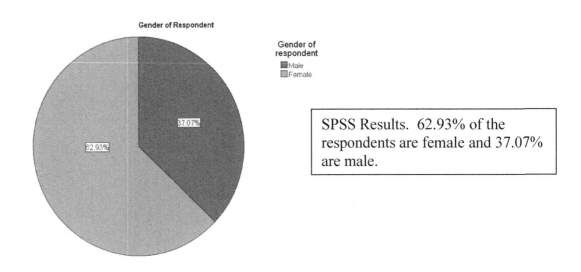

SPSS Results. 62.93% of the respondents are female and 37.07% are male.

Histogram

*The histogram is used primarily to display **<u>interval or scale level</u>** data. A histogram is a series of columns. Each column encompasses a range of values (intervals). The height of the column depicts the number (or percentage) of cases in that interval. A histogram resembles a bar graph except that <u>adjacent columns touch</u> to indicate the continuous nature of the scale.*

Let's generate a histogram of AGE. Use the same procedures as given above except change the chart type to histogram.

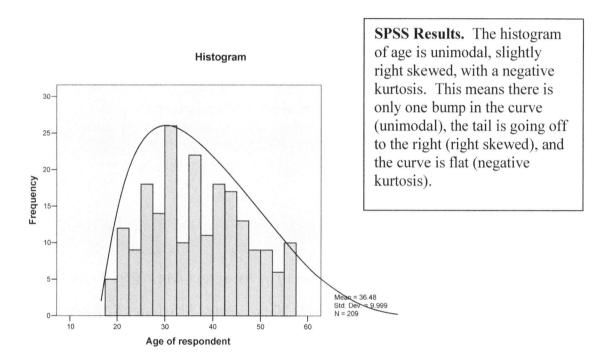

SPSS Results. The histogram of age is unimodal, slightly right skewed, with a negative kurtosis. This means there is only one bump in the curve (unimodal), the tail is going off to the right (right skewed), and the curve is flat (negative kurtosis).

What to Look for in Graphed Distributions

1. Central Tendency	What is the typical or average score, which is indicated by the peak of the array? Is there more than one peak in the distribution? If you wanted to balance the distribution (like a teeter-totter), where would you position the fulcrum (balancing point). This point is the mean.
2. Variability	Do all the cases tend to group together or are they widely scattered?
3. Shape, kurtosis	Does the distribution appear relatively flat or peaked?
4. Shape, skewness	Does the distribution appear symmetrical or lopsided.

- *A distribution is symmetric if the right and left sides of the histogram are approximately mirror images of each other.*
- *It is skewed to the right if the right side of the histogram extends much farther out than the left side (also called Positive Skew).*
- *It is skewed to the left if left side extends further out than the right (also called a Negative Skew).*
- *If the distribution is symmetrical, the values of the mean, median, and mode will be very similar.*
- *In skewed distributions, the mean is pulled towards the "tail" (i.e., the narrow section of the distribution where the extreme scores are located). In skewed distributions, it is better to report the median rather than the mean, as it is a more accurate description of the center of the distribution.*

DETAILS ON SPSS SCREENS FOR GENERATING FREQUENCY CHARTS, DESCRIPTIVE STATISTICS, AND PLOTS.

1. To access the statistical functions of SPSS, select "Analyze" from the Menu Bar.

2. From the Drop-down Menu, select "Descriptive Statistics". This will open another drop-down menu.

3. From this menu, select "Frequencies" which will open another window labeled "Frequencies".

4. From the large box on the left side of the screen, select the variable(s)-of-interest. (Note, more than one variable can be analyzed at the same time).

5. Move these variables to the box labeled "Variables" by clicking the arrow key.

6. Note the three buttons at the bottom of the window, click on each one and note the options available to you.

7. Once you have selected the options that you want SPSS to perform, click "OK".

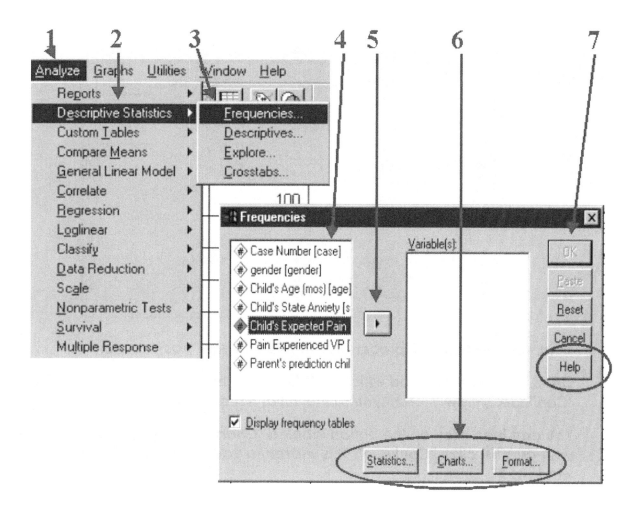

If you click on the "Statistics" button, another window will open labeled "Frequencies Statistics". This window is divided into four sections:

- Percentile Values
- Central Tendency
- Dispersion (or Variability)
- Distribution

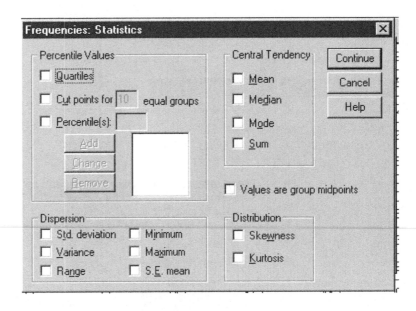

A number of statistical procedures are listed under each section.

Once you have decided which statistical procedures you want to perform, click "Continue". This will return you to the initial "Frequencies" window.

You can then click on the button labeled "Charts". If you have already obtained your graphs, click "Continue". This will return you to the initial "Frequencies" window.

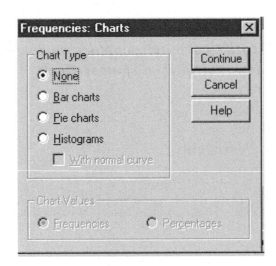

Finally, you can click on the button labeled "Format". This opens another window, which contains information about how the output will be generated.

When you've finished with this window, select "Continue" which will return you to the initial window.

Once you have decided which statistical procedures you want to perform, click "Continue". This will return you to the initial "Frequencies" window.

Chapter 4

Graphs

Goals:

4. Creating Graphs in SPSS
    a. Bar Charts
    b. Pie Charts
    c. Scatter plots
    d. Histograms
5. Using Graph or Chart options to modify a Graph or Chart.

Graphing Data

One of the most effective ways to get to "know your data" is to produce graph. As the old saying goes, a picture is worth a thousand words.

To access the graph options in SPSS, select "Graphs then Legacy Dialogs" from the Menu Bar.

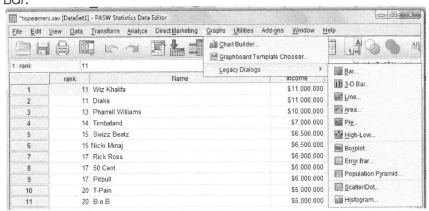

Data Set: topearners.sav, is a collection of data from Forbes.com. The data contains the top 20 Hip-Hop earners in 2011, the top 10 celebrity earners in 2012, and the top 20 athletes from June 2011 – June 2012.

Bar Graphs

Bar graphs are a common way to graphically display the data that represent the <u>frequency of each level of a variable</u>. Here are some examples of when you might use a bar chart:

- Number of voters by political party
- Number of students per section
- Percentage of people voting for Mayoral candidates in San Diego

We will use the top earners data set , topearners.sav, to make the following example bar chart of number of top earners per industry. Follow the steps after the graph to make this graph.

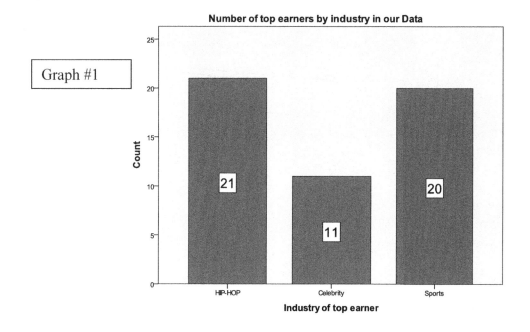

Number of top earners by industry in our Data

Graph #1

Count

21   11   20

HIP-HOP   Celebrity   Sports

Industry of top earner

Steps to Make a Bar Chart in SPSS

1) Open the data file called Chp 5 data.sav.
2) Select **the menu command Graphs > Legacy Dialogs>Bar...**

This will produce the following dialog box:

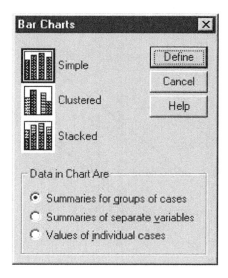

In this dialog box, you must choose between the three types of bar graphs offered by SPSS. The *Simple* bar graph is the most common one and is used to graph frequencies of levels within a variable. In the *Simple* bar graph, the category axis is the variable and each bar represents a level of the variable.

3) Leave Simple Bar chart and Summaries for Groups of Cases selected, and click on Define

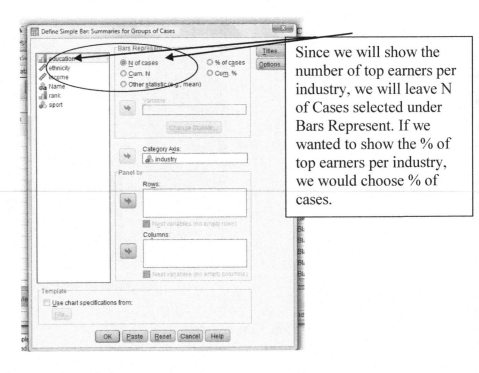

IMPORTANT! This is where you choose to show percent or frequencies (n). If I ask you to graph the percentage of celebrities per industry—you choose percentage here.

Since we will show the number of top earners per industry, we will leave N of Cases selected under Bars Represent. If we wanted to show the % of top earners per industry, we would choose % of cases.

4) Place INDUSTRY in the category axis box
5) We will add a Title to our Chart by clicking the Titles button on the bottom right side of the pop up box shown. Type in the title shown in the box below. Press Continue, and Choose Ok.

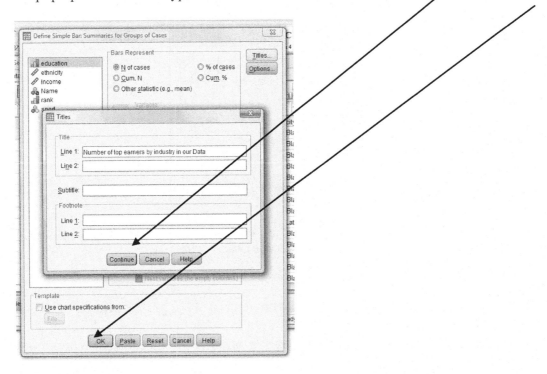

To change the colors of our bars…

Double click on the chart in SPSS to go into the Chart editor window (note the top left corner now says chart editor, and your original graph in the output window is shaded in gray). To change the color or our bars, double click on the bars, they should all be outlined in yellow, then Choose the icon shown in the picture and choose your color choice or in the properties box, choose your fill color and click apply.

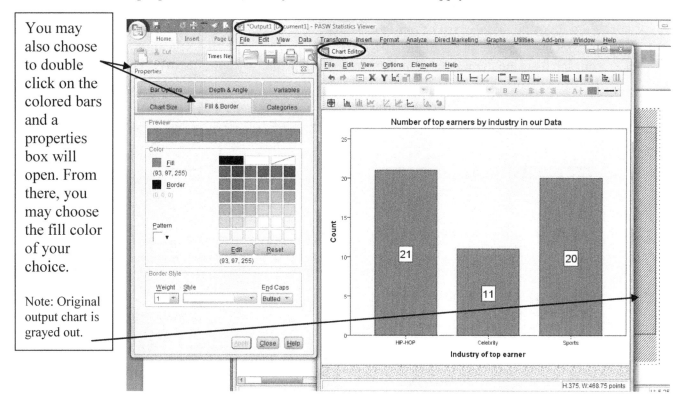

You may also choose to double click on the colored bars and a properties box will open. From there, you may choose the fill color of your choice.

Note: Original output chart is grayed out.

Adding Data Labels, and Changing the Text Font Size of your labels.

Still within the chart editor window, we will add labels to our bars. Go to the menu command Elements and choose Show Data Labels or Choose the corresponding icon (3 blue bars)to do the same thing.

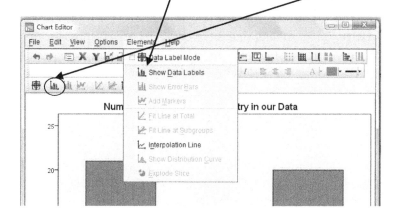

To move your labels to the top of the bar, you can use the Properties box, which opened when you choose to show your data labels. (as shown below) We will choose custom for our Label

Position, and the icon Above Center, so our labels will appear above the bars. It is possible also to drag and drop your label boxes where you want them to be.

We will also make the text of our labels larger. Choose the text tab at the top of the screen, make the font size 12. See picture below. Choose Apply. Additionally, you could change the font of your labels using the menu bar in the chart editor window (default is AUTO, change this to 14 or the size of your choice).

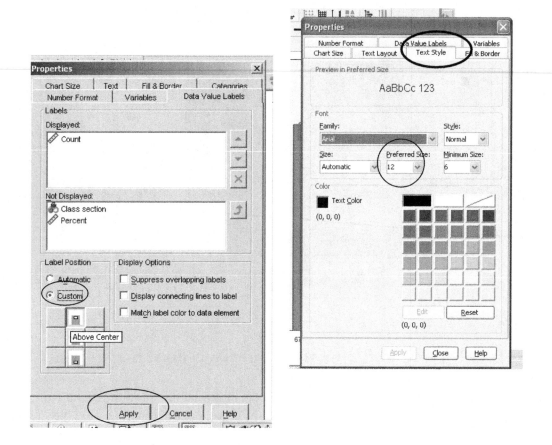

The *Clustered* Bar Chart option shows means that have come from different groups within a certain category. For example, if we wanted to look at film preferences by gender, perhaps comparing those who preferred Batman Returns to Pretty in Pink. A clustered bar chart would show a bar for each gender and those preferring each movie.

For our top earners data, we might look at the education levels of top earners by industry.

To create this clustered bar chart go to the menu command: Graphs>Legacy Dialogs>Bar> choose Clustered and summaries for groups of cases, click Define> place EDUCATION in the category variable box and place INDUSTRY in the Define cluster by box> and click ok.

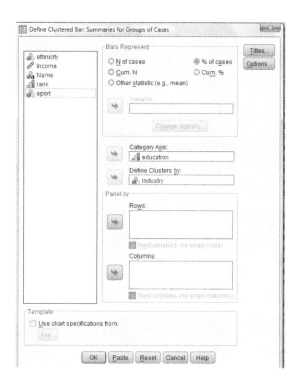

The resulting graph is shown below.

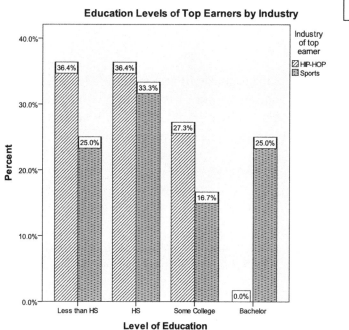

Education Levels of Top Earners by Industry

Graph #2

The other bar graph option for two variables is the **Stacked** option, which will have a bar for each level of one of the variables, while the levels of the other variable are placed on top of each other within each bar. The frequency of each level within a variable in a stacked bar graph is represented by a different color. For example, you could create a graph in which there was a bar for the levels of education, and the percentage of each within Industry type are represented by a different color within a bar:

Graphs>Legacy Dialog>Bar>Stacked >Define> place EDUCATION in Define stacks by box & place INDUSTRY in category axis box>choose bars represent %>ok. Note: We choose to show percentages here rather than counts.

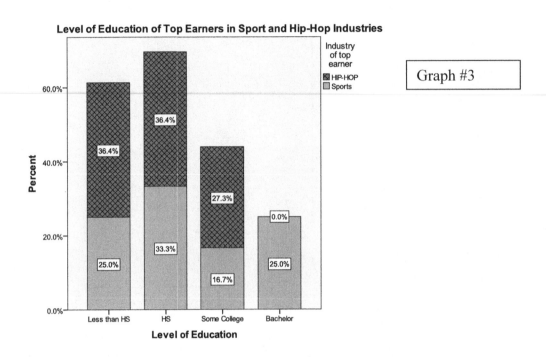

Level of Education of Top Earners in Sport and Hip-Hop Industries

Graph #3

Let's discuss the options in more detail. This discussion will focus on the *Simple* graph. The options for all three are similar, so an understanding of the *Simple* option will suffice for either the *Clustered* or *Stacked* options. To get started with the bar graph, click on the icon representing the type of graph that you want, then click on the **Define** button to produce the following dialog box:

In this dialog box, the *Category Axis* is the only box that you must fill in. Click on a variable name in the list of variables at left and move it to the *Category Axis* box by clicking on the arrow next to that box.

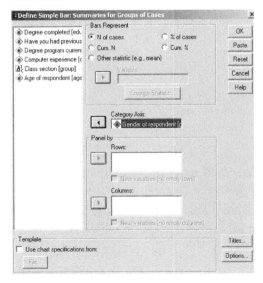

Select the summary information that you want in the *Bars Represent* section of the box.

1.   *N of cases* option is the default and is the most common way of summarizing data in a bar graph.

2.   *% of cases* - There is little difference between using the *N of cases* and the *% of cases* options, other than the unit of measurement in the vertical axis.

3.   Using either the *Cum. n of cases* or the *Cum. % of cases* options will produce a cumulative graph in which the first bar represents the level of a variable with the fewest number of cases, the next bar represents the

level with the second fewest number of cases in addition to the level that has previously been graphed, and so on.

4. You can also use the *Other summary function* option to specify another summary statistic, such as a variable's mean, sum, or variance. As in the above example, begin by selecting a *Category Axis* variable. Next, select the *Other summary function*, then place another variable in the box labels *Variable*.

For example, you could select the variable *INCOME* to obtain the mean income for each level of the variable *INDUSTRY*. Doing so would produce the following table:

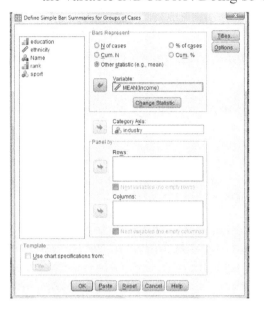

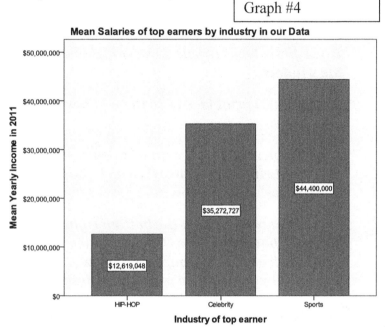

Graph #4

Pie Chart

Pie charts also allow you to compare categories, variables, or individual data values. Pie Chart implies that summary statistics can be regarded as parts of a whole, so counts, percentages and sums are the most commonly used statistics.

To make a pie chart: Graph>Legacy Dialog>Pie>Summaries for groups of cases>Define

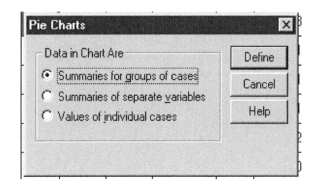

A. Summaries for Groups of Cases - Categories of a single variable are summarized.
B. Summaries of Separate Variables - Several variables are summarized. Each slice represents one of the variables.

C. Values of Individual Cases - A single variable is summarized. Each slice represents the value of an individual case.

Once you click "Define", a new window will open labeled "Define Pie: Summaries for Groups of Cases".

1. The first step is to select the variable-of-interest from the large box on the left side of the window.

2. Move this variable over to the window labeled "Define Slices by" by clicking on the arrow key.

3. Next, you need to indicate what each slice in the pie is to represent. In most cases, you will select either "N of cases" (which counts the number of cases with a specific value) or "% of cases" (which shows the percentage of cases with a specific value).

4. Note the two boxes at the bottom right corner of the window. These boxes are optional, but useful. The "Options" is useful if you have any missing data for the variable. If you click on this button, a window will open which permits you to decide whether you want a slice of the pie to represent missing cases. Often it is easier to interpret the pie if you opt to exclude missing cases. If you want to show your missing data, simply check the box that reads "Display groups defined by Missing Values". Click "Continue" to return to the previous screen.

5. Clicking on the button labeled "Title", permits you to insert a title on to the Pie Chart.

6. Once you have defined the various options, click the button labeled "OK". SPSS will then generate your Pie Chart in the Output window.

Graph #5

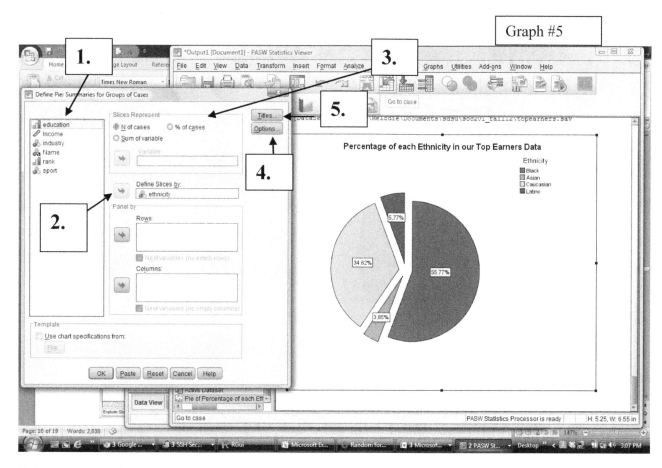

For our data, choose ETHNICITY as the define slices by variable. Ok.

To Explode your pie, meaning separate the pieces of pie, perhaps to show distinction to one or more categories, go into the chart editor, by double clicking your graph in the output window. Then choose the "Pac-man" looking icon – with the smart text "Explode Pie".

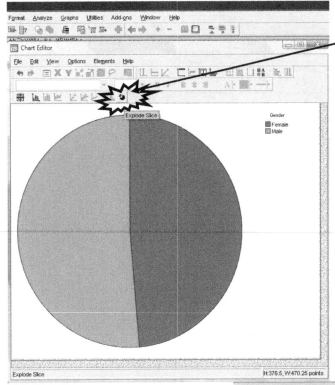

Editing Figures

SPSS permits you to edit the graphs produced (e.g., change colors, labels, or titles). To edit the Pie Chart, simply double-click over the Pie Chart. This will cause another window to open which contains a smaller version of the chart.

Frequently used features:

1. Change Color of a Slice of Pie

Move cursor over the slice of pie that you want to change, and then click the mouse twice, until just the section you want to change is outlined with a yellow line. Next, click on the picture icon of the "Fill".

A color palette will appear that permits you to change the default color. Select your color of choice.

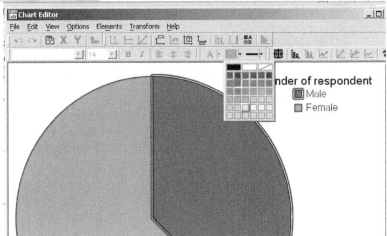

2. Add or Change Labels

To add labels, use Elements>Show Data Labels or the 3 blue bar Icon (as described in the bar chart section).

To change your labels, in the chart editor window, select your pie chart labels by clicking once (the pie labels should be outlined in a yellow line). Then go to Edit>Properties, as shown.

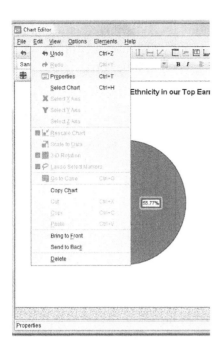

A pop up box will appear. You select the Data Value Labels tab. You can choose to display Ethnicity (which will attach the label Asian, Black, etc. to each piece of pie), the count, which is the N of each pie slice, and/or Percent, which shows the percentage of each category. Simply select the item you want to display in the Not Displayed box and arrow them into the Displayed box.

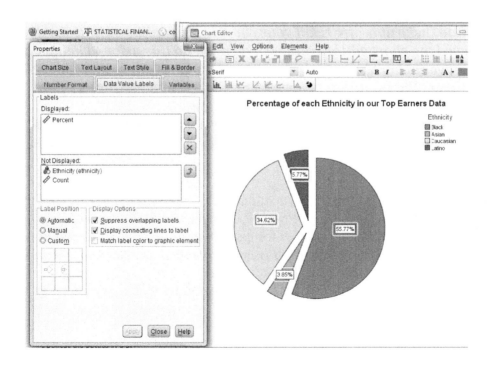

The Label Position area indicates how you would like your pie labels displayed, automatic, or custom with the choice of inside or outside the pie slices. You may also add a connecting line to your labels and pie slices by choosing the Display connecting lines to label box.

Explore the various options, and see how your Pie Chart changes in appearance. You can have SPSS:

- Edit Text (to change the label headings)
- Display Value and/or Percentage of cases represented by a slice.
- Change the color and or fill pattern of your pie slices.

3. Add or Edit Title, Textbox or Footnotes, Hide Legend

Within the Chart Editor, from the Menu-Bar, select "OPTIONS". This will open a drop-down menu from which you can select either the "Title" or "Footnote" or "Textbox". Insert the desired text. There is also a selection to Hide Legend.

Explore the various options and see how you can improve on the appearance of your Pie Chart. Once you have finished editing your graph, simply close the Edit Window.

Moving the graph to another software program file.

- *Single-click on the graph in the "Output" window. This will cause a box to appear around the graph. (NOTE if your chart is gray in the output window, you have not closed the chart editor window. You must do this prior to being able to move or print your chart.)*
- *Select "Edit" from the Menu-Bar*
- *Select "Copy" or "Cut" from the Drop-Down Window.*
- *Open the other program file (i.e., Microsoft Word) and insert the graph by selecting "Edit" from the Menu-Bar, and then "Paste" from the Drop-Down Window.*

SPSS also offer the Export Output option. Export Output saves Viewer output in HTML, text, Word/RTF, Excel, PowerPoint (requires PowerPoint 97 or later), and PDF formats. Charts can also be exported in a number of different graphics formats. *Note: Export to PowerPoint is available only on Windows operating systems and is not available in the Student Version.*

To Export Output   *Make the Viewer the active window (click anywhere in the window).*

From the menus choose: File>Export...   Enter a filename and select an export format. You can export all objects in the Viewer, all visible objects, or only selected objects.

See the SPSS Online Help (menu command HELP) and search for topic Export for more information.

Histogram - *EXAMPLE - Income*

A histogram is a plot showing how many times each score or data point occurs (a graph of the frequency of each data point). Histograms are used with Continuous or Scale variables.

To make a histogram: *Graph>Legacy Dialogs>Histogram>Select the variable-of-interest INCOME>click on the box labeled "Display Normal Curve">add a title>ok. SPSS will generate the histogram.*

Graph #6

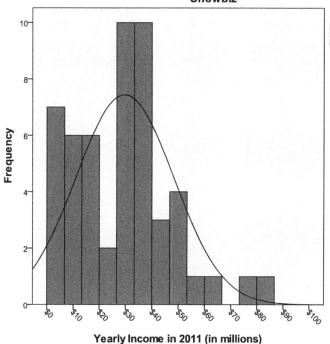

Mean = $29,634,615.38
Std. Dev. = $18,595,918.639
N = 52

Note that I made the following edits to my chart: 1) To edit the x-axis label, I double clicked on the axis label Yearly Income in 2011 in the chart editor, and then added the words (in millions). 2) To format the x-axis, I double clicked the labels and made the following changes in the properties box : a) under the scale tab, I made my major increment 10000000 rather than 20000000; b) Under the tab number format, I made the scaling factor 1000000 and choose scientific notation (never) to show the yearly income in millions of dollars (this makes it easier to read for the end user); c)Under the tab labels and ticks, I choose to display my labels diagonally (label orientation = diagonal).

Scatterplots

Scatterplots give you a tool for visualizing the relationship between two or more variables. Scatterplots are especially useful when you are examining the relationship between continuous variables using statistical techniques such as correlation or regression. To obtain a scatterplot in SPSS, go to the *Graphs* menu and select *Scatter*:

Graphs >Legacy Dialogs>
  Scatter/dot...

This will produce the following dialog box:

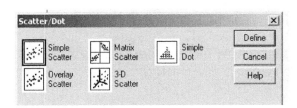

Note that there are five scatterplot options.

1.   The *Simple* option graphs the relationship between two variables.
2.   The *Matrix* option is for two or more variables that you want graphed in every combination: variable is plotted with every other variable. Every combination is plotted twice so that each variable appears on both the X and Y-axis. For example, if you specified a *Matrix* scatterplot with three variables, *Current salary*, *Beginning Salary*, and *Months since hired*, you would receive the following scatterplot matrix:

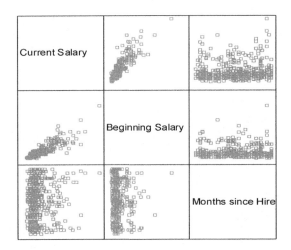

3.   The third scatterplot option is the *Overlay* option. It allows you to plot two scatterplots on top of each other. The plots are distinguished by color on the overlaid plot.
4.   The fourth option for scatterplots is the 3-D scatterplot. This is used to plot three variables in three dimensional spaces.
5.   The fifth option for scatterplots is the simple dot scatterplot. This is used to plot individual observations.

Returning to the *Simple* scatterplot option, you can examine some of the options that are commonly used with a basic scatterplot by plotting the values of two variables.

Since scatterplots require 2 continuous variables, and our top earners data only contains one continuous variable (income), we must open some new data.

OPEN Chp 5 scatterplot data.sav

When you select the *Simple* option from the initial dialog box, you will get the following dialog box:

Select the two variables that you want to plot from the list on the left, placing one in the *Y-axis* box and the other in the *X-axis* box. If you have multiple groups (e.g., males and females) in your

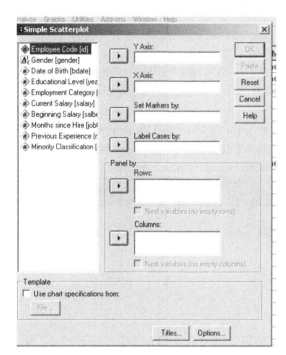

dataset, you can have SPSS draw different colored markers for each group by entering a group variable in the *Set Markers by* box. You can add titles to your chart by clicking on the **Titles** button. We will use Beginning Salary as our X axis variable (SALBEGIN) and Current Salary as our Y axis variable (SALARY) and GENDER as our grouping variable in the Set Markers by box Clicking **OK** will produce the following scatterplot:

Some of the most useful options for modifying your scatterplot are only available after you have the initial scatterplot created. To use these options, double-click on the chart. This will open the chart in a new window.

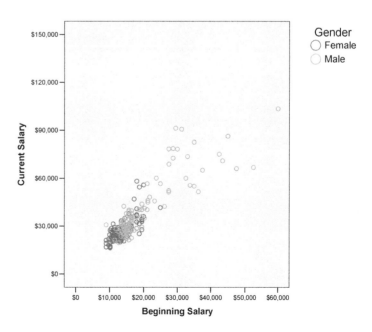

Chapter 5

One Sample T-test

Goals:

1. Given a claim, identify the null hypothesis and the alternative hypothesis and express them.
2. Given a claim and sample data, calculate the t-value or test statistic and p-value.
3. State the conclusions of the hypothesis test in simple, non-technical terms.
4. Calculate the 95% confidence interval, and state the interval in non-technical terms.

The DATA   You can find this data in Blackboard under Course Documents, Data Sets for SPSS Lab Book. The data is called one sample t data.sav. The data is the mean body temperature of 106 healthy adults as found by SDSU researchers. There is only one variable in the data, as follows:

Variable	Description
BODYTEMP	Body temperature in degrees Farenheit

Hypothesis Testing

In this chapter, we will use a scale variable or a continuous variable, and compare the mean of our scale variable to a known population mean.

> In Statistics, a **hypothesis** is a claim or statement about a property (such as the mean) of a population.

OUR EXAMPLE PROBLEM

Medical researchers claim that the mean body temperature of healthy adults is different than 98.6 degrees Fahrenheit.

Level of Significance (Alpha)

For our class, we will use an alpha value of .05. Level of Significance (Alpha) reflects the willingness of the researcher to make a wrong conclusion. To incorrectly reject a "true" null hypothesis (Type I error).

A Type I error is made when a researcher reports that a difference (or a relationship) exist when in fact it does not.

> **Traditionally, alpha is set at .05.** *Using this value, the researcher is indicating a willingness to make a wrong conclusion (in rejecting the Null Hypothesis) five times out of 100.*
>
> *Different alpha values can be set depending on the nature of the study:*
>
> - *If potentially serious consequences could occur if a wrong decision is made, a researcher may opt to set a more stringent level of significance may be set.* **For example, an alpha of .01 indicates a willingness to make an error once in a 100 times.**
> - *Conversely, if a research study is exploratory in nature, has few associated risks, and a small sample size is used, a researcher may opt to reduce alpha by setting it at .10.* **This value indicates a willingness to make a wrong conclusion 10 times in a 100.**

> *Statistical Significant Findings:* During data analysis if the calculated probability (p-value) is **less than** the predefined alpha (i.e., a value less than .05); you reject the Null Hypothesis and conclude that the sample is not representative of the population. In other words, you reject the idea that chance alone produced the observed findings (i.e., the Null Hypothesis). You report a statistically significant finding
>
> *Nonsignificant findings* indicate that the results are **not too** unlikely to have occurred due to sampling error, alone. Similar differences might be found even if no real relationship exists in the population. When the observed value is greater than .05, the Null Hypothesis is not rejected.

It is important to appreciate that failure to reject the null hypothesis does not prove that a "real" relationship or difference does not exist in the population. *Failure to reject the null only indicates that the data were not sufficient to detect it.*

Results of Statistical Test	Interpretation
Calculated p-value < alpha	**Reject null hypothesis of no difference or no relationship**
Calculated p-value > alpha	**Fail-to-Reject null hypothesis of no difference or no relationship**

Decide whether to use a One or Two-tailed test

In a two-tailed statistical inference tests, the specified "region of rejection" (alpha = .05) is divided into two. Half (2.5% or .025) the region of rejection is located in the lower tail and half in the upper tail. Using a two-tailed test, the researcher would reject the Null hypothesis if calculated scores were either too high or too low (either extreme).

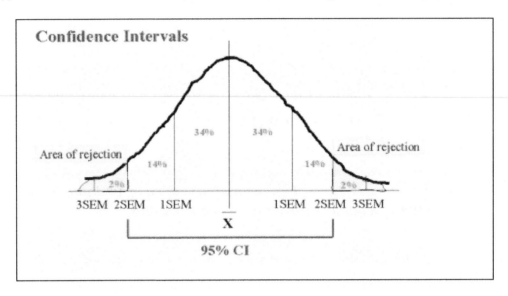

Two-tailed tests indicate that both tails of a sampling distribution are used to determine the critical region for rejecting the null hypothesis.

*Look what happens to the diagram if you opt to use a **one-tailed test**. The entire region of rejection is located in one end of the distribution. By positioning the entire rejection of rejection (.05) in one tail of the sampling distribution, you increase the probability of obtaining a statistically significant finding - **increased statistical power**.*

When to Use One-Tailed versus Two-Tailed Tests

An advantage of a one-tailed test is that it increases the likelihood of achieving statistically significant results if scores act as predicted. Using a one-tailed test is comparable to doubling the odds that you will obtain a significant finding.

The drawback is that you can't switch the region of rejection after the fact. Once you have committed to a one-tailed test, you must stick with it regardless of the consequences.

Because of this, a one-tailed test should be used only if there is a convincing reason for predicting the direction of the relationship or difference.

Steps in Hypothesis Testing

Step 1. State your Null and Alternative Research Hypothesis

- *A research hypothesis is a statement predicting the nature (and possibly the direction) of the proposed relationship or difference between variables or groups.*

- A hypothesis test actually includes two statistical hypotheses - the Null Hypothesis and its Research Alternative.
- **The Null Hypothesis** (Ho) is a statement about the value of the population parameter (like the mean), and it must contain the equals sign (=). This is the hypothesis that we generally hope to reject at the end of our analysis of the data.
  - There is **no difference** between the sample mean and the population mean. Any observed differences can be attributed to chance (sampling error) alone.
- **The Alternative Hypothesis** (Ha) is a statement about the value of the population parameter (like the mean), and it must contain the not equals sign ($\neq$), a greater than sign (>) or a less than sign (<). If you use a not equal sign, you are conducting a 2 tailed test. If you use the > or < sign in your alternative hypothesis, you will be conducting a 1 tailed test. This is the hypothesis that we generally hope to prove at the end of our analysis of the data.

Null Hypothesis:

For our example, our Null Hypothesis is as follows:

Ho: The mean body temperature is 98.6 degrees Fahrenheit.
(**OR** Ho: Mean body temperature = 98.6 degrees Fahrenheit.)

Alternative Hypothesis

For our example, our Alternative Hypothesis is as follows:

Ha: The mean body temperature is not 98.6 degrees Fahrenheit.
(**OR** Ha: Mean body temperature $\neq$ 98.6 degrees Fahrenheit.)

Step 2. Assumptions

Assumptions of the One Sample t-test

- *Sample randomly selected*
- *Dependent variable is a continuous variable measured at the interval or ratio level*
- *Mean score for population is known*
- *Sample data are approximately normally distributed*
- *Note: Test is quite robust to minor violations of assumptions of normal distributed data when sample sizes $\geq$ 30*

We must examine our data to determine if we meet our assumptions.

The first assumption is the data is randomly selected. We will assume this is true, since we can't test it with SPSS, we must rely that the researchers at SDSU collected the data randomly.

The second assumption is that our dependent variable is continuous. Our variable is bodytemp. It is continuous and a scale variable.

The third assumption is that the mean of the population is known. For our data set, the population mean is 98.6 degrees Fahrenheit.

Finally, we must test that the sample data are normally distributed. We will use SPSS to test this assumption.

Analyze>Descriptive Statistics>Explore>put bodytemp in dependent list, click statistics button and check descriptives, continue, ok. Then, select plots, check histogram and normality plots with test> continue, ok.

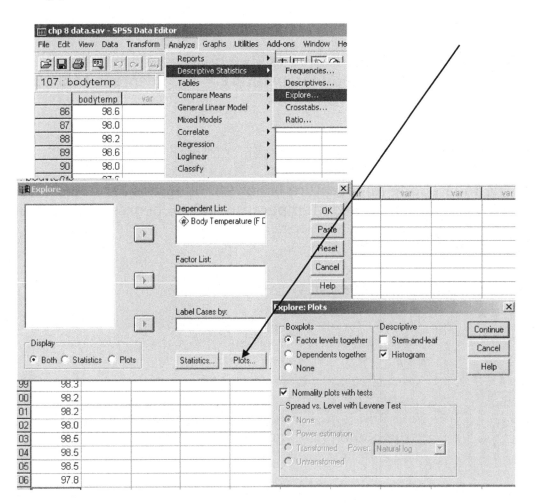

Is the data normal? *We could look at the histogram. If the histogram has a bell shaped curve, then we would say it looks normal. For our data, the peak is higher than expected for normal distribution, also left skewed. The histogram indicates the data is not normal. IF this happens, we would not use the t-test, because the assumption for the t-test that the data be approximately normal is not met. However, due to a large sample size (n=106), we can use the Central Limit Theorem to ensure the sampling distribution of the mean is approximately normal. That's what the note at the bottom of the assumptions box refers to.* **(Note: Test is quite robust to minor violations of assumptions of normal distributed data when sample size ≥ 30)**

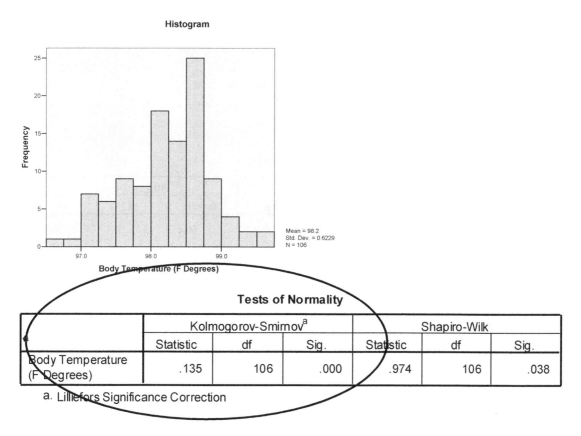

Histogram

Tests of Normality

	Kolmogorov-Smirnov[a]			Shapiro-Wilk		
	Statistic	df	Sig.	Statistic	df	Sig.
Body Temperature (F Degrees)	.135	106	.000	.974	106	.038

a. Lilliefors Significance Correction

I believe a better approach to testing normality is using the Kolmogorov Smirnov test. This table is produced in your output.

First, we will adjust the cell width of the Sig column by double clicking on the table in SPSS output window. Click once on the .000 column under Sig. and then right click and choose Cell Properties. Choose the fifth option under format value, the scientific notation option. See pictures below, and hit apply.

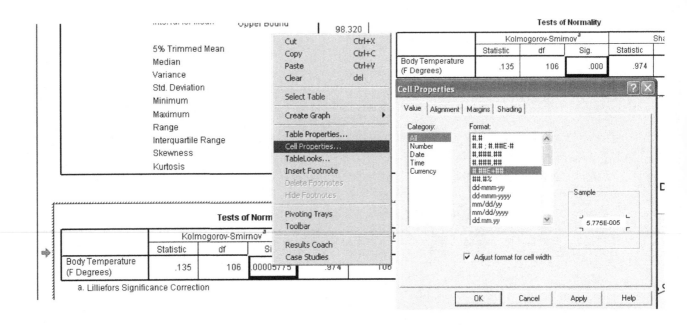

Tests of Normality

	Kolmogorov-Smirnov[a]			Shapiro-Wilk		
	Statistic	df	Sig.	Statistic	df	Sig.
Body Temperature (F Degrees)	.135	106	********	.974	106	.038

a. Lilliefors Significance Correction

Double click on the test of normality table above in SPSS. Move your cursor over the line to the right of Sig. until you get a double-sided arrow, and then stretch the box to the right until you can view the entire number not ****.

Your new table should look like this:

Tests of Normality

	Kolmogorov-Smirnov[a]			Shapiro-Wilk		
	Statistic	df	Sig.	Statistic	df	Sig.
Body Temperature (F Degrees)	.135	106	5.775E-005	.974	106	.038

a. Lilliefors Significance Correction

Remember, we are testing to see if the data is normally distributed. We must complete this before we go on to our t-test.

We have an Ho and an Ha for the test of normality.

Ho: The data is normally distributed.   Ha: The data is not normally distributed.

Kolmogorov-smirnov value (K/S) = .135   p-value=.00005775 (5.77 ^ -5)

79

Conclusion: We reject ho, our data is not normally distributed. Our p-value is less than .05. But we can continue with our t-test because we have a large sample size (larger than 30), n=106.

Remember the RULE, if our p-value is less than alpha (.05), we reject ho and if our p-value is larger than alpha (.05), we fail to reject ho.

MEAN

This table produced in your output shows the 95% confidence interval of the mean. We will use this later in our analysis. Also, we have the sample mean, 98.2 degrees. We are testing to see if 98.2 is significantly different from 98.6 degrees with 95% confidence.

Descriptives

			Statistic	St
Body Temperature (F Degrees)	Mean		98.200	
	95% Confidence Interval for Mean	Lower Bound	98.080	
		Upper Bound	98.320	
	5% Trimmed Mean		98.206	
	Median		98.400	
	Variance		.388	
	Std. Deviation		.6229	
	Minimum		96.5	
	Maximum		99.6	
	Range		3.1	
	Interquartile Range		.8	
	Skewness		-.336	
	Kurtosis		-.205	

Confidence Interval

Case Processing Summary

	Cases					
	Valid		Missing		Total	
	N	Percent	N	Percent	N	Percent
Body Temperature (F Degrees)	106	100.0%	0	.0%	106	100.0%

HINT – If you don't know what your sample size is, you can find it here in the case processing summary in your SPSS output. Our n=106.

Step 3. Calculate t-value and p-value

Deciding whether to reject or accept the Null Hypothesis is determined by calculating our t-value (test statistic).

The formula for calculating the test statistic for a one sample t-test is simply:

$$t\text{-}value \ = \ \frac{\textit{Sample Mean} - \textit{Population Mean}}{\textit{Standard Deviation} \ / \ (\sqrt{n})}$$

$$'t = (98.2\text{-}98.6)/0.06050 = -6.611$$

Note: *Standard Deviation / ($\sqrt{n}$) = SEM*

The significance of this "**calculated t-value**" is determined by comparing it to a predefined "critical t-value".

This critical value is determined from a "**theoretical sampling distribution of t-values**". An interesting characteristic of the "sampling distribution of t" is that the shape changes, depending on the size of the samples being drawn. With sample sizes less than 50, the t-distribution tends to be flatter and have fatter tails than a normal distribution. However with samples greater than 50, the t-distribution is approximately normal in shape. The following diagram illustrates how the shape of the t-distribution varies depending on the sample size.

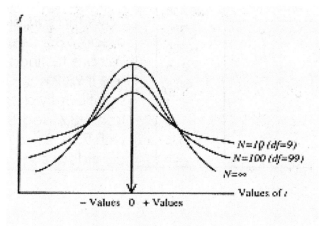

Sampling Distribution of t-values with Varying Sample Sizes

Two factors determine the "critical value" for a particular statistical analysis:

- Level of significance (alpha)
- Degrees of Freedom (df)

Degrees of Freedom (df) is an awkward statistical concept to get a handle on. However for our purposes, it is sufficient to understand that it is closely linked to the sample size. In fact, the formula for calculating df in a one sample t-test is simply:

$$\text{Degrees of Freedom} = \ n - 1$$

$$n = \text{Sample size}$$

Df for our example = 105

Although SPSS will do this work for you, it is useful to understand how df is determined for the common statistical tests. With this knowledge, you can check the accuracy of the analysis. If the df reported on the SPSS output doesn't look right, you may have made a mistake in entering the analysis command.

When it performs an analysis, SPSS automatically compares the calculated t-value to the critical t-value, and then it generates a probability value indicating the likelihood that these two mean scores came from the same population.

Interpretation

- **If the calculated statistical value is greater than the critical statistical value reject the Null hypothesis.**
- **If the calculated probability (p-value) is less than the predefined alpha value (usually set at .05), reject the Null Hypothesis as a statistically significant result exists.**

To obtain a One-Sample T-test in SPSS

Analyze>Compare Means>One-SampleT-test>

Select the variable you want to test (bodytemp)

Enter your population mean (this is the number in your null hypothesis) into the test value box (98.6)

Click on options to change the confidence level (1-99 are your choices; we will keep our confidence level at 95%).

OK.

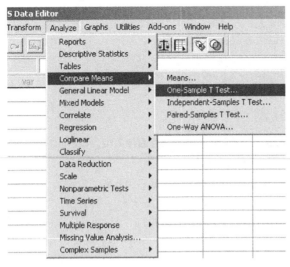

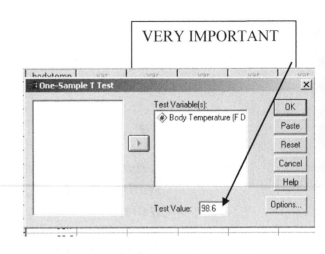

VERY IMPORTANT

Make sure this circle contains your population mean, the number you are testing, contained in Ho.

ORIGINAL SPSS OUTPUT

One-Sample Test

	Test Value = 98.6					
					95% Confidence Interval of the Difference	
	t	df	Sig. (2-tailed)	Mean Difference	Lower	Upper
Body Temperature (F Degrees)	-6.611	105	.000	-.4000	-.520	-.280

Edited SPSS Output

One-Sample Test

	Test Value = 98.6					
					95% Confidence Interval of the Difference	
	t	df	Sig. (2-tailed)	Mean Difference	Lower	Upper
Body Temperature (F Degrees)	-6.611	105	.000000002	-.4000	-.520	-.280

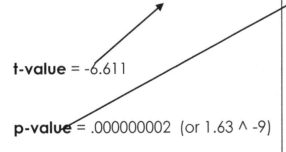

t-value = -6.611

p-value = .000000002 (or $1.63 \wedge -9$)

How did I get the edited output above, which shows the actual p-value (Sig 2-tailed)? First, we will adjust the cell width of the Sig column by double clicking on the table in SPSS output window. Click once on the .000 column under Sig. and then right click and choose Cell Properties. Choose scientific notation or choose to increase your decimal places to 9.

Step 4. Conclusions

Conclusion in Words: Since our p-value is smaller than .05; we reject ho and conclude that the mean body temperature of healthy adults is not 98.6 degrees Fahrenheit.

Since we are reasonably confident the average body temperature of healthy adults is not 98.6 degrees, what is it? We can calculate a range of values – an interval – that should include the population mean 95% of the time.

95% confidence interval

We will use our original descriptive box for our confidence interval.

To obtain the 95% confidence interval for the one-sample t-test in SPSS:

Analyze>Descriptive Statistics>Explore>put bodytemp in dependent list, click statistics button and check descriptives, continue, ok.

Descriptives

			Statistic	Std. E
Body Temperature (F Degrees)	Mean		98.200	.0
	95% Confidence Interval for Mean	Lower Bound	98.080	
		Upper Bound	98.320	
	5% Trimmed Mean		98.206	
	Median		98.400	
	Variance		.388	
	Std. Deviation		.6229	
	Minimum		96.5	
	Maximum		99.6	
	Range		3.1	
	Interquartile Range		.8	
	Skewness		-.336	.235
	Kurtosis		-.205	.465

95% confidence interval of the mean

We are 95% confident that the interval 98.08 to 98.32 contains the true average body temperature in degrees Fahrenheit of healthy adults.

What is the relationship between hypothesis tests and confidence intervals?

If a value is not contained in the 95% confidence interval (98.6 in our example), we reject the hypothesis that it is a plausible value for the population mean, using a 5% criterion.

We check for the population mean in our confidence interval, if it is in the interval, we should FAIL TO REJECT Ho.

If the population mean is not in our confidence interval, we should REJECT Ho.

Since 98.6 is not in the interval (98.08, 98.32), we should REJECT Ho, which is what we did!

Let's review the key steps:

1. We started by stating our Ho and Ha, using our population mean.
   Ho: Mean body temperature = 98.6 degrees Fahrenheit
   Ha: Mean body temperature ≠98.6 degrees Fahrenheit

2. We tested our assumption of normality, for our continuous variable, bodytemp.
   Ho: The data is normally distributed.
   Ha: The data is not normally distributed.

   Kolmogorov-smirnov value (K/S) = .135   p-value=.00005775

   Conclusion: We rejected ho, our data was not normally distributed. Our p-value is less than .05. But we continued with our t-test because we had a large sample size (larger than 30), n=106.

3. We found our t-value and p-value with SPSS, and stated each
   t-value = -6.611   **p-value** = .000000002 (or 2 ^ -9)

4. We stated our conclusions, based on our t-value and p-value.
   <u>**Conclusion in Words:**</u> Since our p-value is smaller than .05, we reject ho and conclude that the mean body temperature of healthy adults is not 98.6 degrees Fahrenheit.

5. We found our 95% confidence interval.

   We are 95% confident that the interval 98.08 to 98.32 contains the true average body temperature in degrees Fahrenheit of healthy adults.
   a. We checked our confidence interval for 98.6. Since 98.6 was not in the interval, we should have rejected ho, which is what we did!

Our key points:

- The common belief was that the population mean was 98.6.
- We calculated a sample mean of 98.2.
- Considering the distribution of sample means, the sample size, and the magnitude of the difference between 98.2 and 98.6, we found a sample mean of 98.2 was unlikely to occur if the population mean is really 98.6.
- There are 2 possible explanations for the sample mean of 98.2:
    - Either a very rare event occurred OR
    - The population mean is not 98.6

- Since our probability of getting a sample mean of 98.2 (when the population mean is 98.6) is so low, we go with the more reasonable choice that the value of the population mean is not 98.6 as is commonly believed. Thus, we rejected Ho.

Example 2: Test if the mean # of siblings for average Americans was 2 in 1991. We will use the data set One sample t example 2.sav found in Blackboard. This data set is the U.S. General Social Survey for 1991.

Our main variable of interest is sibs.

Step 1. State your Null and Alternative Research Hypothesis

Ho: The mean # of siblings for the average American in 1991 was 2.

Ha: The mean # of siblings for the average American in 1991 was not 2.

Step 2. Assumptions

The first assumption is the data is randomly selected. We will assume this is true.

The second assumption is that our dependent variable is continuous. Our variable is sibs. It is continuous and a scale variable.

The third assumption is that the mean of the population is known. For our data example, we will assume the population mean is 2.

Finally, we must test that the sample data are normally distributed.

State ho and ha for the test of normality:

Ho: The data is normally distributed.

Ha: The data is not normally distributed.

Tests of Normality

	Kolmogorov-Smirnov[a]			Shapiro-Wilk		
	Statistic	df	Sig.	Statistic	df	Sig.
Number of Brothers and Sisters	.176	1505	2.53E-130	.881	1505	.000

a. Lilliefors Significance Correction

Since our pvalue = 2.53^{-130} is much less than .05, we reject ho and conclude the data is not normal. However, we can continue with our t-test because we have a large sample size of 1505 people.

Step 3. Calculate t-value and p-value

One-Sample Test

	Test Value = 2					
					95% Confidence Interval of the Difference	
	t	df	Sig. (2-tailed)	Mean Difference	Lower	Upper
Number of Brothers and Sisters	24.596	1504	1.5E-112	1.932	1.78	2.09

Step 4. Conclusions

Reject ho. Since our pvalue is less than .05, we reject ho and conclude that the mean # of siblings for the average American in 1991 was not 2.

95% confidence interval: (3.78, 4.09)

We are 95% confident that the true mean # of siblings for the average American in 1991 was between 3.78 and 4.09.

Descriptives

			Statistic	Std. Error
Number of Brothers and Sisters	Mean		3.93	.079
	95% Confidence Interval for Mean	Lower Bound	3.78	
		Upper Bound	4.09	
	5% Trimmed Mean		3.69	
	Median		3.00	
	Variance		9.282	
	Std. Deviation		3.047	
	Minimum		0	
	Maximum		26	
	Range		26	
	Interquartile Range		3	
	Skewness		1.468	.063
	Kurtosis		3.507	.126

Chapter 6

Paired Samples T-test

Goals:

5. Given a claim, identify the null hypothesis and the alternative hypothesis and express them.
6. Given a claim and sample data, calculate the t-value or test statistic and p-value.
7. State the conclusions of the hypothesis test in simple, non-technical terms.
8. Calculate the 95% confidence interval, and state the interval in non-technical terms.
9. Explore non-parametric options for the paired sample t-test.

In chapter 5, we discussed the t-test procedure for a one sample t-test. The purpose was to determine whether a specific sample mean differed from a hypothesized population mean. In this chapter, we will be making comparisons between 2 groups.

The paired samples t-test is used in 2 situations. The first type of comparison is made between 2 scores that belong to the same group. For example, a pre-post repeated measures experiment where individuals are measured twice, before and after a treatment. For instance, if you were asked to evaluate your SPSS anxiety at the first week of class and at the end of the semester, a paired t-test would be used because the evaluations are not independent; you made both evaluations, so they are dependent.

The second type of comparison can be made between two related samples on the same dependent variable. Suppose we asked husbands and wives to respond to a measure of marital happiness. Since husbands and wives are related, we would keep their responses together during the analysis, using a paired t-test.

Other examples of paired t-tests:

- Suppose you are testing a weight loss drug. You would measure weight before the study and after the study (after they took the pill for 4 weeks). We would see if there is a significant difference in their weight before and after taking the weight loss pill.

- **Crest and Dependent Samples**
  In the late 1950s, Procter & Gamble introduced Crest toothpaste as the first such product with fluoride. To test the effectiveness of Crest in reducing cavities, researchers conducted experiments with several sets of twins. One of the twins in each set was given Crest with fluoride, while the other twin continued to use ordinary toothpaste without fluoride. It was believed that each pair of twins would have similar eating, brushing, and genetic characteristics. Results showed that the

twins who used Crest had significantly fewer cavities than those who did not. This use of twins as dependent samples allowed the researchers to control many of the different variables affecting cavities.

Example 1 Using SPSS

Do female students minimize their weight?

Female students were given a survey that included a question asking them to report their weight in pounds. They weren't told that their weight would be measured, but weights were accurately measured after the survey was completed. Anonymity was maintained using ID numbers, so no personal information would be announced. Is there a significant difference between the reported weights and the measured weights?

The DATA You can find this data in Blackboard under Course Documents, Data Sets for SPSS Lab book. The data is called Pairedt data.sav. The data is actual and reported weights of 51 females students as found by SDSU researchers. The variables in the data, are as follows:

Variable	Description
ID	Student Id
Repweight	Weight as reported by student (in Pounds)
Actualweight	Actual weight of student as measured by researchers (in Pounds)
Diff	The difference between the reported and actual weight of the students (Reported – Actual weight)

Steps in Hypothesis Testing

Step 1. State your Null and Alternative Research Hypothesis
- **The Null Hypothesis** (Ho) states there is **no difference** between the two group scores and it must contain the equals sign (=). *This is the hypothesis that we generally hope to reject at the end of our analysis of the data.*

- *The group scores are **the same or equal**. Any observed differences can be attributed to chance (sampling error) alone.*

- **The Alternative Hypothesis** (Ha) states there is **a difference** between the two group scores, and it must contain the not equals sign (≠), a greater than sign (>) or a less than sign (<). If you use a not equal sign, you are conducting a 2 tailed test. If you use the > or < sign in your alternative hypothesis, you will be conducting a 1 tailed test. *This is the hypothesis that we generally hope to prove at the end of our analysis of the data.*

- *The group scores are **not the same or not equal**. Any observed differences cannot be attributed to chance (sampling error) alone.*

Null Hypothesis:

For our example, our Null Hypothesis is as follows:

Ho: There is **no difference** between the actual and reported weights of female students.
> (**OR** Ho: Female students actual and reported weights are **the same** or equal. Reported weight **equals** actual weight of female students)

Alternative Hypothesis

For our example, our Alternative Hypothesis is as follows:

Ha: There is **a difference** between the actual and reported weights of female students.
> (**OR** Ha: Female students actual and reported weights are **not the same** or not equal. Reported weight **does not equal** actual weight of female students)

Step 2. Assumptions

Assumptions of the Paired Samples t-test

- *Sample randomly selected*
- *Dependent variable is a continuous variable measured at the interval or ratio level*
- *Data are approximately normally distributed*

- *Note: Test is quite robust to minor violations of assumptions of normal distributed data when sample sizes ≥ 30*

We must examine our data to determine if we meet our assumptions.

The first assumption is the data is randomly selected. We will assume this is true, since we can't test it with SPSS, we must rely that the researchers at SDSU collected the data randomly.

The second assumption is that our dependent variable is continuous. Our variables, ACTUALWEIGHT and REPWEIGHT, are continuous and thus scale variables.

Finally, we must test that the sample data are normally distributed. We will use SPSS to test this assumption.

To test Normality, we must use the variable DIFF, which is the difference between the actual and reported weights. In the example data, I have computed DIFF for you. This is how to compute DIFF, if there is no difference variable in your data set.

In the Data View window, go to

TRANSFORM>COMPUTE

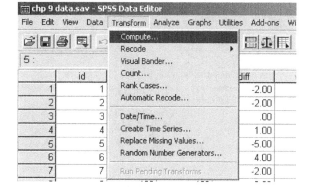

Type in your new variable or Target variable name (DIFF), you may give your variable a Label by clicking on the Type & Label button.

In the numeric expression box, arrow in the "reported weight" click the minus key in the calculator area, and arrow in the "actual weight". We are just telling SPSS to calculate the difference between the actual and reported weight.

Click OK. (If SPSS asks you if you want to replace the existing variable, click OK) You will have created a new variable, which is the difference between our measurements.

See picture on the next page.

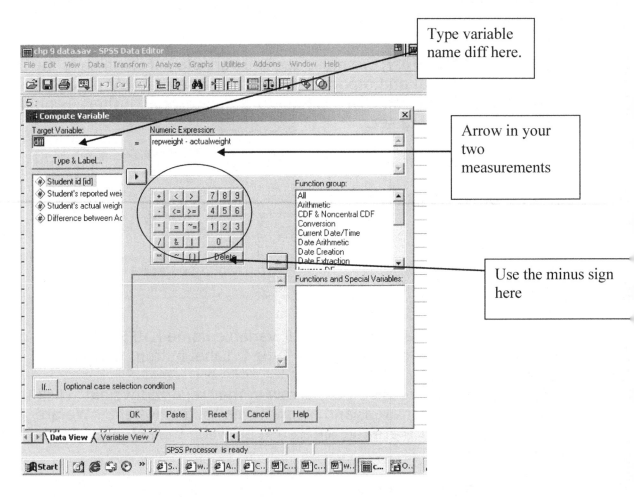

Once you have the variable DIFF in your data set, go to

Analyze>Descriptive Statistics>Explore>put DIFF in dependent list, click statistics button and check descriptives, continue, ok. Then, select plots, check normality plots with test> continue, ok.

This Test of Normality can be found in your output, as shown below.

Tests of Normality

For instructions to adjust the width of this cell, see Chp 8.

	Kolmogorov-Smirnov[a]			Shapiro-Wilk		
	Statistic	df	Sig.	Statistic	df	Sig.
Difference between Actual and Reported Weights	.179	51	.00031465	.949	51	.029

a. Lilliefors Significance Correction

Null and Alternative hypothesis for Testing Assumption of Normality
Ho: The data is normally distributed.
Ha: The data is not normally distributed.

Our k/s value = .179   **Our p-value** = .00031465

Conclusion for Test of Normality: Reject ho, our data is not normally distributed. Our p-value is smaller than .05 (alpha), so we reject ho. But since we have a large sample size, larger than 30, we can continue with our paired t-test.

Remember the RULE, if our p-value is less than alpha (.05), we reject ho and if our p-value is larger than alpha (.05), we fail to reject ho.

You can find your sample size in your output, see the Case Processing Summary and note the number shown under Valid, N, that is our sample size. For our data, we have 51 female students. Since we have a sample size larger than 30, we can continue with the t-test even though our data is not normal.

Case Processing Summary

	Cases					
	Valid		Missing		Total	
	N	Percent	N	Percent	N	Percent
Difference between Actual and Reported Weights	51	100.0%	0	.0%	51	100.0%

Step 3. Calculate t-value and p-value

Deciding whether to reject or accept the Null Hypothesis is determined by calculating our t-value (test statistic) and p-value.

Computing the Paired-Samples T-test

Analyze>Compare Means>Paired-Samples > Select REPWEIGHT then ACTUALWEIGHT>move the pair into the paired variables box>ok.

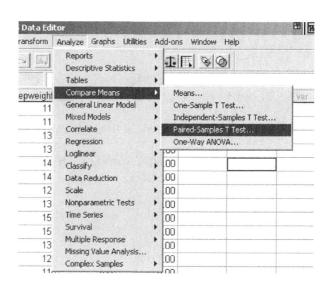

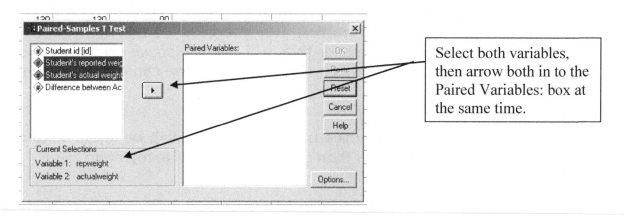

Select both variables, then arrow both in to the Paired Variables: box at the same time.

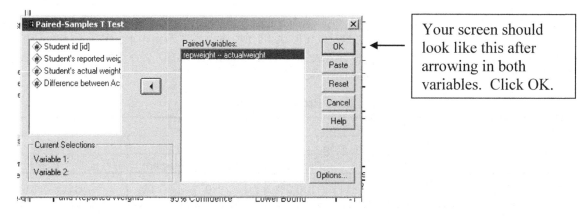

Your screen should look like this after arrowing in both variables. Click OK.

Paired Samples Statistics

		Mean	N	Std. Deviation	Std. Error Mean
Pair 1	Student's reported weight	127.20	51	16.412	2.298
	Student's actual weight	127.41	51	14.837	2.078

The box above should appear first in your output. This box tells us the sample mean of each variable.

The Student's mean reported weight was 127.20 pounds
The Student's mean actual weight was 127.41 pounds.

We are trying to determine if these two means are statistically the same or different.

We now look at the Paired Samples T-test box in our output, to find the t-value and p-value, which will lead us to our conclusion.

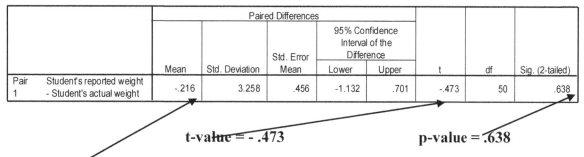

Paired Samples Test

		Paired Differences							
					95% Confidence Interval of the Difference				
		Mean	Std. Deviation	Std. Error Mean	Lower	Upper	t	df	Sig. (2-tailed)
Pair 1	Student's reported weight - Student's actual weight	-.216	3.258	.456	-1.132	.701	-.473	50	.638

t-value = - .473   p-value = .638

The mean difference in the actual and reported weights is -.216 pounds.

Step 4. Conclusions

*Our p-value is larger than .05 (alpha), therefore we **fail to reject ho**.*

Conclusion in Words: Since our p-value is larger than .05; we fail to reject ho and conclude that the actual weight and reported weight of female students is the same. Female students do not minimize their weight.

We can calculate a range of values – an interval – that should include the population mean 95% of the time.

95% confidence interval

We will use our Paired Samples t-test box to find our confidence interval.

Paired Samples Test

		Paired Differences							
					95% Confidence Interval of the Difference				
		Mean	Std. Deviation	Std. Error Mean	Lower	Upper	t	df	Sig. (2-tailed)
Pair 1	Student's reported weight - Student's actual weight	-.216	3.258	.456	-1.132	.701	-.473	50	.638

We are 95% confident that the interval -1.132 to 0.701 contains the true average difference in the actual and reported body weights (in pounds) of female students.

What is the relationship between hypothesis tests and confidence intervals?

We are now looking for the value of 0 in our confidence interval. If there is no difference in the actual and reported weights of female students, then subtracting the 2 weights would give a value of 0. So we are saying in Ho, there is no difference in the actual and reported weights, the difference between them is 0. So if 0 is in our confidence interval, we fail to reject ho. If 0 is not in our interval, we reject the null hypothesis, using a 5% criterion.

We check for 0 in our confidence interval, if it is in the interval, we should FAIL TO REJECT Ho.

If the value 0 is not in our confidence interval, we should REJECT Ho.

Since 0 is in the interval (-1.1, .7), we should FAIL TO REJECT Ho, which is what we did!

Let's review the key steps:

1. We started by stating our Ho and Ha.
   Ho: There is **no difference** between the actual and reported weights of female students.
   Ha: There is **a difference** between the actual and reported weights of female students

2. We tested our assumption of normality, for our continuous variable, DIFF.
   Ho: The data is normally distributed.
   Ha: The data is not normally distributed.

   Kolmogorov-smirnov value (K/S) = .179   p-value=.0003

   Conclusion: We rejected ho, our data was not normally distributed. Our p-value is less than .05. But we continued with our t-test because we had a large sample size (larger than 30), n=51.

3. We found our t-value and p-value with SPSS, and stated each
   t-value = -.473   **p-value** = .638

4. We stated our conclusions, based on our t-value and p-value.
   <u>**Conclusion in Words:**</u> Since our p-value is larger than .05; we fail to reject ho and conclude that the actual weight and reported weight of female students is the same.

Our key points:

- The common belief was that female students minimized their weight.
- We calculated the sample mean of their reported weight of 127.2 and the sample mean of their actual weight of 127.41 pounds. The mean difference was -.216. We were looking to see if -.216 is significantly different from 0 (ho – there is no difference).
- Considering the distribution of sample means, the sample size, and the magnitude of the difference, we found a sample mean difference of -.216 was likely to occur if there was no difference in the actual and reported weights.
- Since our probability of getting a sample mean difference of -.216 (when the population mean difference is 0) is so high, we say there is insufficient evidence to suggest a difference in the actual and reported weights. Thus, we failed to reject Ho.

Example 2

In 1987, 3 researchers investigated the possible role of β-endorphins in the collapse of runners. β-endorphins are morphine-like substances manufactured in the body. They measured plasma β-endorphins concentrations for 11 runners before and after they participated in half marathon runs. The question of interest was whether or not β-endorphins levels changed during a run. The researchers hypothesized that runners were able to continue running despite pain and discomfort because β-endorphin levels increased and produced a sense of well-being. Since β-endorphin level was measured twice on each subject; this is an example of a paired t-test.

The DATA You can find this data in Blackboard. The data is called pairedt data example 2.sav. The data is the β-endorphin levels of 11 runners before and after a half marathon. The variables in the data, are as follows:

Variable	Description
BEFORE	β-endorphin level before the half marathon run
AFTER	β-endorphin level after the half marathon run
Diff	**The difference between the before and after measurements of the runners (After - Before)**

Research Hypotheses

Step 1. State your Null and Alternative Research Hypothesis

Null Hypothesis

There is no difference between the average β-endorphins levels for the runners.

(In general, the null hypothesis for a paired design is there is no difference between the average values for the 2 members of a pair in a population. This is the same as testing whether the two means are equal.)

Alternative Hypothesis

There is a difference between the average β-endorphins levels for the runners.

(In general, the alternative hypothesis is there is a difference in the average values.)

Step 2: Null and Alternative hypothesis for Testing Assumption of Normality

Ho: The data is normally distributed.
Ha: The data is not normally distributed.

Our k/s value = .129   **Our p-value** = .200

Tests of Normality

	Kolmogorov-Smirnov[a]			Shapiro-Wilk		
	Statistic	df	Sig.	Statistic	df	Sig.
DIFF	.129	11	.200*	.969	11	.872

*. This is a lower bound of the true significance.

a. Lilliefors Significance Correction

> Since our p-value is greater than alpha, our data is normal.

Conclusion for Test of Normality: Since our p-value is greater than .05 (alpha), we fail to reject ho and conclude the data is normally distributed.

STEP 3 Calculate the t-value and p-value for the paired samples t-test.

Analyze>Compare Means>Paired-Samples > Select after then before (results in after-before)>move the pair into the paired variables box>ok.

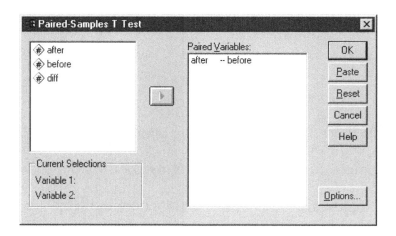

Paired Samples Statistics

		Mean	N	Std. Deviation	Std. Error Mean
Pair 1	before	8.4273	11	4.24832	1.28092
	after	27.1636	11	9.67794	2.91801

These are the mean beta endorphin levels before and after the run.

t value=7.460

p-value is 2.16 ^ -5 which is less than alpha (.05); you can edit the cell to find the exact value p-value.

Paired Samples Test

		Paired Differences						
				Std. Error Mean	95% Confidence Interval of the Difference			
		Mean	Std. Deviation		Lower	Upper	t	df	Sig. (2-tailed)
Pair 1	before - after	-18.73636	8.32974	2.51151	-24.33236	-13.14037	-7.460	10	.000

Average difference between before and after marathon values is 18.74

Step 4. Conclusions

<u>*Conclusion:*</u> *Reject ho, our p-value is smaller than .05. We can conclude that β-endorphin levels appear to change during a half-marathon run.*

95% confidence interval of the difference: (-24,332, -13.140)
95% confidence interval of the difference in words: We are 95% confident that the true mean difference in the Beta endorphin levels of runners before and after a half-marathon run is between -24, and -13 units.

We are double checking for 0 in the interval (as 0 means no difference or Ho is true).
Since 0 is not in our interval (-24,-13); we should have rejected ho, which is what we did.

Chapter 7

Independent Samples T-test

Goals:

10. Given a claim, identify the null hypothesis and the alternative hypothesis and express them.
11. Given a claim and sample data, calculate the t-value or test statistic and p-value.
12. State the conclusions of the hypothesis test in simple, non-technical terms.
13. Calculate the 95% confidence interval, and state the interval in non-technical terms.

In chapter 6, we discussed the t-test procedure for paired samples. The purpose was to determine whether two paired sample means were the same or different statistically. The sample means were related, such as measurements taken twice on the same person (before and after) or measurements taken on related subjects, twins or husbands and wives. In this chapter, we will be making comparisons between 2 groups that are not related, thus independent.

> *Independent (or between subjects) t-test* compares the means scores obtained from two unrelated sample groups to determine if they are significantly different.

Example 1 Using SPSS

Do new textbooks cost more or less at SDSU than at UCSD?

Students were asked to report their average cost of their textbooks at UCSD and SDSU.

The DATA You can find this data in Blackboard under Course Documents, Data Sets for SPSS Lab book. The data is called indept.sav. The data is the reported cost of the average textbook of 68 students, 40 UCSD students and 28 SDSU students. The variables in the data, are as follows:

Variable	Description
school	1=UCSD or 2=SDSU
newbook	Price of average new textbook

Steps in Hypothesis Testing

Step 1. State your Null and Alternative Research Hypothesis

- **The Null Hypothesis** (Ho) states there is **no difference** between the two group scores and it must contain the equals sign (=). This is the hypothesis that we generally hope to reject at the end of our analysis of the data.

- The group scores are **the same or equal**. Any observed differences can be attributed to chance (sampling error) alone.

- **The Alternative Hypothesis** (Ha) states there is **a difference** between the two group scores, and it must contain the not equals sign (≠), a greater than sign (>) or a less than sign (<). If you use a not equal sign, you are conducting a 2 tailed test. If you use the > or < sign in your alternative hypothesis, you will be conducting a 1 tailed test. This is the hypothesis that we generally hope to prove at the end of our analysis of the data.

- The group scores are **not the same or not equal**. Any observed differences cannot be attributed to chance (sampling error) alone.

Null Hypothesis:
For our example, our Null Hypothesis is as follows:

Ho: There is **no difference** between the average price of new textbooks at UCSD and SDSU.
> (**OR**   Ho: New textbook prices are **the same** or equal at UCSD and SDSU. SDSU textbook price **equals** UCSD textbook price)

Alternative Hypothesis
For our example, our Alternative Hypothesis is as follows:

Ha: There is **a difference** between the average price of new textbooks at UCSD and SDSU.
> (**OR**   Ha: New textbook prices are **not the same** or not equal at UCSD and SDSU. SDSU textbook price **does not equal** UCSD textbook price)

STEP 2 Assumptions

> ### *Assumptions of the Independent t-test*
>
> *Samples randomly selected from population & Group membership determined by random assignment if constructed groups.*
>
> *Dependent variable is a continuous variable measured at the interval or ratio level*
>
> *Two levels of the independent variable*
>
> *Data from the two sample groups are approximately normally distributed*
>
> *Data from the two sample groups have approximately the same variance (i.e., homogeneity of variance)*
>
> ***Note: When sample sizes are large (i.e., when both groups have > 25 subjects) and are approximately equal in size, this test is quite robust to violations of the assumptions of normalcy and homogeneity of variance***

We must examine our data to determine if we meet our assumptions.

The first assumption is the data is randomly selected. We will assume this is true, since we can't test it with SPSS, we must rely that the researchers at SDSU collected the data randomly.

The second assumption is that our dependent variable is continuous. Our variable, NEWBOOK, is continuous and thus a scale variable.

The third assumption is that our independent variable has two levels Our variable, SCHOOL, has two levels, UCSD and SDSU are the two schools we are comparing. This independent variable can also be called a FACTOR or GROUPING Variable.

> ### DEPENDENT VARIABLE = NEWBOOK
> ### INDEPENDENT VARIABLE = SCHOOL

The fourth assumption requires that we test that the sample data are normally distributed. We will use SPSS to test this assumption.

To test Normality,

Analyze>Descriptive Statistics>Explore>put NEWBOOK in Dependent List box and SCHOOL in Factor List box, click statistics button and check descriptives, continue, ok. Then, select plots, check normality plots with test> continue, ok.

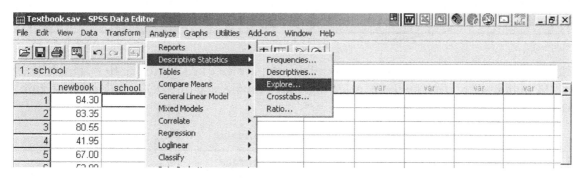

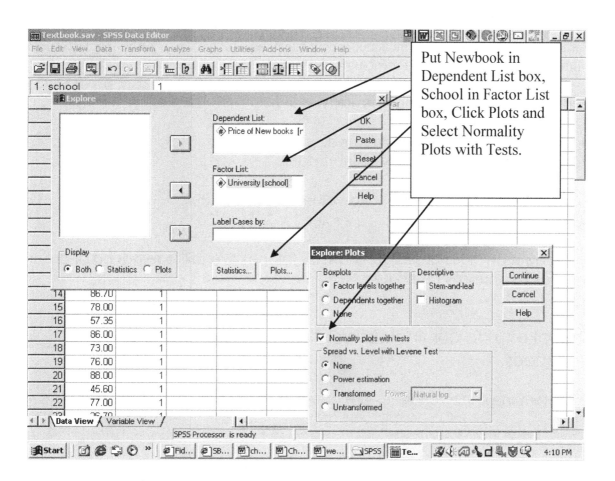

Put Newbook in Dependent List box, School in Factor List box, Click Plots and Select Normality Plots with Tests.

This Test of Normality can be found in your output, as shown below.

Case Processing Summary

		Cases						
		Valid		Missing		Total		
	University	N	Percent	N	Percent	N	Percent	
Price of New books	UCSD	40	100.0%	0	.0%	40	100.0%
	SDSU	28	100.0%	0	.0%	28	100.0%

The Case Processing Summary box shows the sample size of each group, 40 = sample size of UCSD students reporting textbook price and 28 = sample size of SDSU students reporting textbook price.

Tests of Normality

		Kolmogorov-Smirnov[a]			Shapiro-Wilk		
	University	Statistic	df	Sig.	Statistic	df	Sig.
Price of New books	UCSD	.129	40	.093	.946	40	.054
	SDSU	.155	28	.082	.929	28	.057

a. Lilliefors Significance Correction

k/s values

p-values

For the Test of Normality for the independent t-test, we will have two Kolmogorov-Smirnov values and two p-values (one for each group).

Remember, we are testing normality:

Ho: The data is normally distributed.
Ha: The data is not normally distributed.

K/S for UCSD = .129   p-value for UCSD = .093
K/S for SDSU = .155   p-value for SDSU = .082

Since both p-values are larger than .05 (alpha), we fail to reject ho and conclude the data is normally distributed.

This takes us to our final assumption, the assumption of equal variances.

To determine if the variances are equal or not equal, we use SPSS.

Steps in SPSS to determine if the variances are equal.

Analyze>Compare means>Independent Samples T-test>

Our test variable is NEWBOOK (the dependent variable)

Select Grouping variable whose values define the two groups (SCHOOL). Click define groups and indicate how the groups are defined.

> *Select <u>Use specified values</u> – We enter a 1 for UCSD and a 2 for SDSU. This is how our groups are defined, a 1 = UCSD and a 2 = SDSU. These are our values. We can find these values by looking in the data view window and noting the values for school are 1 and 2. See picture below.*

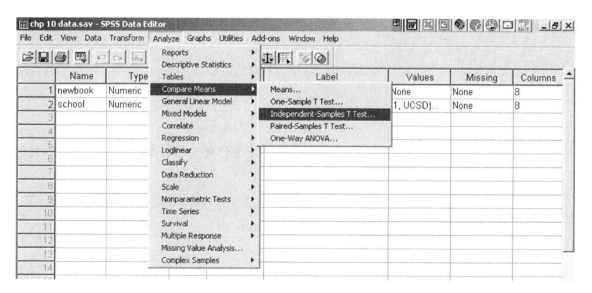

Click ok and OK.

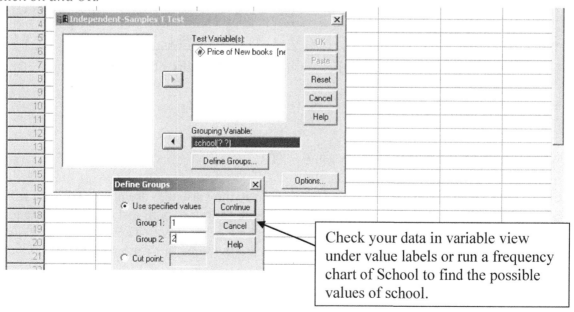

Check your data in variable view under value labels or run a frequency chart of School to find the possible values of school.

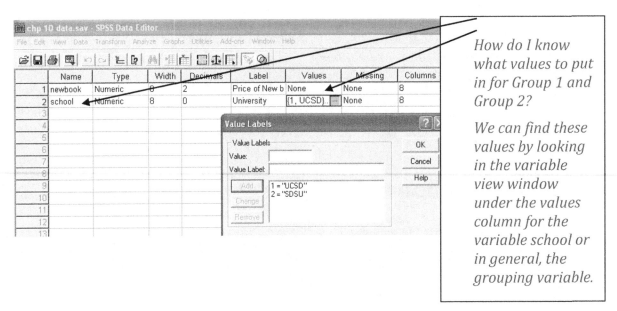

How do I know what values to put in for Group 1 and Group 2?

We can find these values by looking in the variable view window under the values column for the variable school or in general, the grouping variable.

RESULTS OF THE LEVENE'S TEST FOR EQUALITY OF VARIANCES

Independent Samples Test

		Levene's Test for Equality of Variances		t-test for Equality of Means						
									95% Confidence Interval of the Difference	
		F	Sig.	t	df	Sig. (2-tailed)	Mean Difference	Std. Error Difference	Lower	Upper
Price of New books	Equal variances assumed	9.843	.003	1.696	66	.095	8.39964	4.95338	-1.49009	18.28938
	Equal variances not assumed			1.828	65.581	.072	8.39964	4.59598	-.77762	17.57691

The circled portion of the output indicates the Test for Equality of Variances.

Ho: The variances are equal.
Ha: The variances are not equal.

We get an F value = 9.843   and a p-value = .003

Since our p-value is smaller than .05 (alpha), we say the variances are not equal. We will use the bottom row of the output for our t-value and p-value for our overall test of means. Remember, we are trying to test if the price of textbooks are the same or different. The stuff we have done up until now is required prior to testing if the prices are the same or different.

STEP 3 Independent Samples t-test
Obtain t-value and p-value from OUTPUT

Independent Samples Test

		Levene's Test for Equality of Variances		t-test for Equality of Means					95% Confidence Interval of the Difference	
		F	Sig.	t	df	Sig. (2-tailed)	Mean Difference	Std. Error Difference	Lower	Upper
Price of New books	Equal variances assumed	9.843	.003	1.696	66	.095	8.39964	4.95338	-1.49009	18.28938
	Equal variances not assumed			1.828	65.581	.072	8.39964	4.59598	-.77762	17.57691

Use this row to obtain your t-value and your p-value. WHY? B/c our F=9.843 and p=.003 for the Equality of Variance test. We are testing Ho: Variances of the groups are the same. Ha: Variances of the 2 groups are different. Since our p-value is less than 0.05 (alpha), we reject ho and conclude that equal variances are NOT assumed.

$$T=1.828 \qquad p\text{-value} = .072$$

Step 4 Conclusion

Since our p-value is greater than .05, we fail to reject ho and conclude there is no significant difference between the average text book prices at UCSD and SDSU.

Step 5   95% confidence interval

We are 95% confident that the true difference in the average price of textbooks at UCSD and SDSU is between -.78 cents and $17.58.

We are checking for 0 in our confidence interval, because if 0 is in our interval it is like saying there is no difference between the textbook prices. The prices are the same (if we subtracted them, we would get a value of 0).

Since 0 is in our interval, we should fail to reject ho, which is what we did.

Step 6   *Sample means*

Group Statistics

	University	N	Mean	Std. Deviation	Std. Error Mean
Price of New books	UCSD	40	65.1175	23.07569	3.64859
	SDSU	28	56.7179	14.78864	2.79479

The above table is produced in your independent t-test (and equality of variances test) output. It shows the mean price of textbooks at UCSD is $65.12 and the mean price of textbooks at SDSU is $56.72.

Let's review the key steps:

1. We started by stating our Ho and Ha.

Ho: There is **no difference** between the average price of new textbooks at UCSD and SDSU.

   Ha: There is **a difference** between the average price of new textbooks at UCSD and SDSU.

2. We tested our assumption of normality, for our continuous variable, NEWBOOK, using SCHOOL as our grouping variable, with 1 and 2 defining our groups (1=UCSD and 2 = SDSU).

   Ho: The data is normally distributed.
   Ha: The data is not normally distributed.

   K/S for UCSD = .129   p-value for UCSD = .093
   K/S for SDSU = .155   p-value for SDSU = .082

   Conclusion: We failed to reject ho, our data was normally distributed. Our p-values were greater than .05.

3. We tested for equality of variances.
   Ho: The variances are equal.
   Ha: The variances are not equal.

   We get an F value = 9.843   and a p-value = .003

   Therefore, the variances were not equal; our p-value was less than .05, so we used the bottom row of our t-table output.

4. We found our t-value and p-value with SPSS, and stated each
   t-value =1.828   **p-value** = .072

5.We stated our conclusions, based on our t-value and p-value.

Conclusion in Words: Since our p-value is larger than .05; we fail to reject ho and conclude there is no significant difference between the average text book prices at UCSD and SDSU.

Our key points:

- The common belief was that textbooks at UCSD and SDSU were priced differently.
- We calculated the sample mean of price of textbooks at UCSD, $65.12 and the mean price of textbooks at SDSU, $56.72. We were looking to see if these 2 means were significantly different or not.
- Considering the distribution of sample means, the sample size, and the magnitude of the difference, we found a sample mean difference of 8.399 (65.12 – 56.72) was likely to occur if there was no difference in the textbook prices at UCSD and SDSU.
- Since our probability of getting a sample mean difference of 8.399 is so high, we say there is insufficient evidence to suggest a difference in the textbook prices at UCSD and SDSU. Thus, we failed to reject Ho.

Example 2

Suppose we have 2 independent groups of people – those who are satisfied with their jobs and those who are not. We want to determine if the average ages are the same for the two groups.

Remember – independence means there is no relationship between the people in the different groups.

The DATA You can find this data in Blackboard, under Course Documents. The data is called indept example 2.sav. The data is the age of satisfied and unsatisfied employees. There are 325 satisfied and 419 unsatisfied employees. The variables in the data, are as follows:

Variable	Description
age	Employee age
Satjob2	Satisfaction of employees with their job 1 = very satisfied and 2 = not very satisfied

Null Hypothesis: In the population, the average age of very satisfied full-time workers and not very satisfied full time workers is the same.

Alternative Hypothesis: In the population, the average age of very satisfied full-time workers and not very satisfied full time workers is different.

Assumptions: 1) *Calculate the test of normality:*

Analyze>Descriptive Statistics>Explore> select Age as the Dependent Variable and satjob2 as the independent. Select display plots only. Ok.

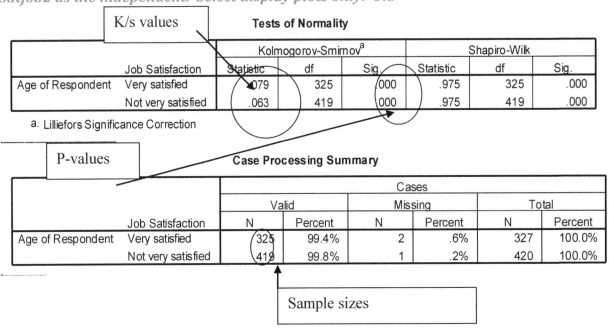

K/s values

Tests of Normality

	Job Satisfaction	Kolmogorov-Smirnov[a]			Shapiro-Wilk		
		Statistic	df	Sig.	Statistic	df	Sig.
Age of Respondent	Very satisfied	.079	325	.000	.975	325	.000
	Not very satisfied	.063	419	.000	.975	419	.000

a. Lilliefors Significance Correction

P-values

Case Processing Summary

	Job Satisfaction	Cases					
		Valid		Missing		Total	
		N	Percent	N	Percent	N	Percent
Age of Respondent	Very satisfied	325	99.4%	2	.6%	327	100.0%
	Not very satisfied	419	99.8%	1	.2%	420	100.0%

Sample sizes

111

Ho: Data is normally distributed   Ha: Data is not normally distributed.

<u>Conclusion for the test of normality:</u> Since both of our p-values are smaller than .05 (alpha), we reject ho and conclude the data is not normally distributed. However, we have large sample sizes, n = 325 in the satisfied group and n=419 in the not satisfied group. Since the sample size in each group is larger than 30, we can continue with the t-test even though the data was not normally distributed.

Recall the Note in the assumptions box: When sample sizes are large (i.e., when both groups have > 30 subjects) and are approximately equal in size, this test is quite robust to violations of the assumptions of normalcy and homogeneity of variance

To obtain the test for equality of variance and independent samples t-test in SPSS:

Analyze>Compare means>Independent Samples T-test>

Select variables whose mean you want to test. (age)

Select variable whose values define the two groups (satjob2). Click define groups and indicate how the groups are defined, 1 = satisfied and 2 = unsatisfied.

Independent Samples Test

		Levene's Test for Equality of Variances		t-test for Equality of Means					95% Confidence Interval of the Difference	
		F	Sig.	t	df	Sig. (2-tailed)	Mean Difference	Std. Error Difference	Lower	Upper
Age of Respondent	Equal variances assumed	.377	.540	2.347	742	.019	1.93	.822	.315	3.544
	Equal variances not assumed			2.327	672.439	.020	1.93	.829	.301	3.558

Results for Levene's test of equal variances

Use this row to obtain your t-value and your p-value. WHY? B/c our F=.377 and p=.540 for the Equality of Variance test.   We are testing Ho: Variances of the groups are the same. Ha: Variances of the 2 groups are different. Since our p-value is greater than 0.05 (alpha), we fail to reject ho and conclude that

t-value = 2.347   p-value = .019

Conclusion: *Since we found that the equal variances assumption was met, we will use the equal variances assumed row, for the observed difference of 1.93 years, the t statistic is 2.35. The observed two-tailed significance level or p-value is 0.019. This means that only 1.9% of the time would you expect to see a sample difference of 1.93 years or larger, if the two population means were equal.*

REJECT HO! *Since .019 is less than .05, we reject the null hypothesis that the satisfied employees and the not satisfied employees have the same average age.*

<u>Confidence Interval for the Mean Difference</u> = (.32, 3.54) *We are 95% confident that the true difference in the mean age of the satisfied and unsatisfied employees is from 0.32 years to 3.54 years.*

Since 0 is not in our interval, we should have rejected ho, which is what we did.

Chapter 8

Cross tabulation and Chi-Square Analysis

Goals:

6. Understand Observed and Expected Counts.
7. Creating cross tabs.
8. Interpreting the Chi-square statistic.

One of the most frequently asked questions in studies is, "Are these 2 variables related?". Is education related to voting behavior? Do more educated people tend to vote democratic?

Is marital status related to happiness? Are married people happier than single people?

If we wanted to know if there is any relationship between the gender of undergraduates at SDSU and their footwear preferences, we might select 50 males and 50 females as randomly as possible, and ask them, "On average, do you prefer to wear sandals, sneakers, leather shoes, boots, or something else?" We might find that males prefer to wear sneakers and females prefer to wear sandals. Our results might show gender is related to foot wear choice or knowing someone's gender will help us to determine their shoe style.

> **Crosstabulation** - To show in tabular form the relationship between 2 or more categorical variables.

If we wanted to look at difference among car colors and region of the country, it doesn't make sense to compare means, because we have 2 categorical variables. What would the mean of region give us, if we have values of West, South, North, and East? Therefore, we use a crosstabulation to look at the percentages of people falling within the categories.

Categorical variables – 2 or more distinct categories

Gender – male or female

Ethnicity – Asian, white, Hispanic

Grade – a, b, c, d, f

We will make a cross tabulation with continuous data only if it is categorized – like age. (0-19, 20-39, etc)

THE DATA we will use for this chapter is called chisq data.sav. It is located in Blackboard under Course Documents, Data sets for SPSS Lab book. This data set contains the voting choices of people in the 1992 Presidential election. **Note: If you are using the student version of SPSS, you will not be able to open this data set as it contains more cases than allowed in the student version of SPSS (1500). You should review this exercise in one of the labs on campus.**

First example:

"Do you think more women voted for President Clinton than voted for President Bush?" Is gender related to the likelihood of voting for Clinton?

Step 1:

State your null hypothesis (We write this as H_o:)

Ho: There is **no relationship** between gender and choice for President. (You might also phrase this as Ho: The **same percentage** of men and women voted for Clinton in 1992. or Ho: Gender and choice for president are **independent**.)

Note the key words above in bold print. Ho always has **no relationship**, the **same percentage** or **independent** as part of its statement. This never changes! Basically, we are writing there is no relationship between variable 1 and variable 2. You just need to plug in the labels for those variables, and you've written your null hypothesis.

Independence means knowing the value of one of these variables for a case tells you nothing about the value of the other variable.

Step 2:

State your alternative hypothesis (We write this as H_a:)

Ha: There is **a relationship** between gender and choice for President. (You might also phrase this as Ha: A **different percentage** of men and women voted for Clinton in 1992. or Ha: Gender and choice for president are **dependent**.)

Note the key words above in bold print. Ha always has **a relationship**, a different **percentage** or **dependent** as part of its statement. <u>This never changes</u>! Basically, we are writing there is a relationship between variable 1 and variable 2. Knowing the value of one variable tells us something about the value of the second variable. You just need to plug in the labels for those variables, and you've written your alternative hypothesis.

Step 3:
Make the cross tabulation table to explore the relationship between our two categorical variables.

First let's open up our data set. chisq data.sav
Then go to

Analyze>Descriptive Statistics>Crosstabs>

Row variable=sex, Column Variable = pres92

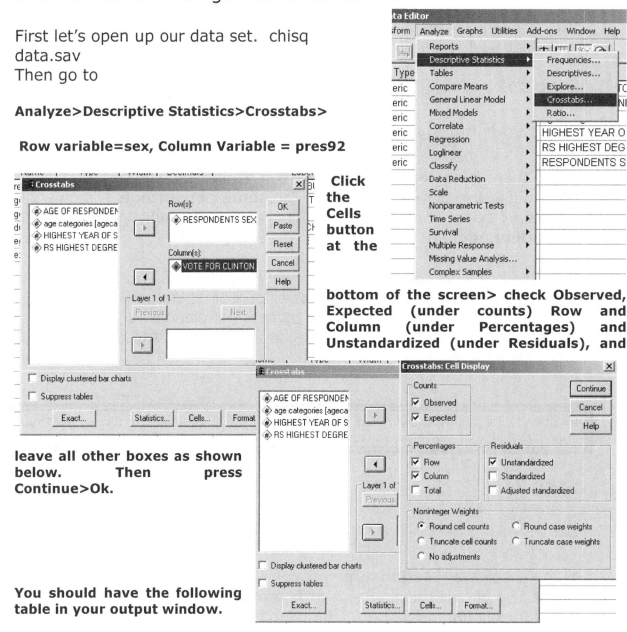

Click the Cells button at the

bottom of the screen> check Observed, Expected (under counts) Row and Column (under Percentages) and Unstandardized (under Residuals), and

leave all other boxes as shown below.   Then   press Continue>Ok.

You should have the following table in your output window.

RESPONDENTS SEX * VOTE FOR CLINTON, BUSH, PEROT Crosstabulation

			VOTE FOR CLINTON, BUSH, PEROT			Total
			Bush	Perot	Clinton	
RESPONDENTS SEX	male	Count	315	152	337	804
		Expected Count	287.7	121.0	395.3	804.0
		% within RESPONDENTS SEX	39.2%	18.9%	41.9%	100.0%
		% within VOTE FOR CLINTON, BUSH, PEROT	47.7%	54.7%	37.1%	43.5%
		Residual	27.3	31.0	-58.3	
	female	Count	346	126	571	1043
		Expected Count	373.3	157.0	512.7	1043.0
		% within RESPONDENTS SEX	33.2%	12.1%	54.7%	100.0%
		% within VOTE FOR CLINTON, BUSH, PEROT	52.3%	45.3%	62.9%	56.5%
		Residual	-27.3	-31.0	58.3	
Total		Count	661	278	908	1847
		Expected Count	661.0	278.0	908.0	1847.0
		% within RESPONDENTS SEX	35.8%	15.1%	49.2%	100.0%
		% within VOTE FOR CLINTON, BUSH, PEROT	100.0%	100.0%	100.0%	100.0%

The basic element of cross tabulation is the count of the number of cases in each cell of the table. The statistical procedure we will use to test the null hypothesis is based on comparing the observed count in each cell to the expected count. Expected count is the number of cases we would expect to find in a cell if the null hypothesis is true. So, if the expected count and the actual count are the same, then ho is most likely true. If the expected count and actual count are very different, then we will reject ho (or state ho is not true) and conclude we have enough evidence to say ha is true.

Residual = Difference between observed and expected counts.

A positive residual means you observed more cases in a cell than you would expect if the null hypothesis were true. A negative residual means you observed fewer cases than you would expect if the null hypothesis were true.

For our particular data, more males voted for Bush than expected 27.3 more. Fewer males voted for Clinton than expected, 58.3 fewer. That is if there was no relationship between gender and who they voted for in 1992.

The sum of the expected counts for any row or column is the same as the observed count for that row or column. Residuals sum to 0 across any row or column.

Row Percentages:

For our data, this is the % within respondent's sex row. The interpretation of this data is as follows, **OF ALL MEN**, 39.2% voted for Bush, 18.9% voted for Perot, and 41.9% voted for Clinton. Notice the % sum to 100%, across the row. (See the squares in the table on the previous page)

Similarly, if I asked you, **OF ALL WOMEN**, what percentage voted for Clinton? You should answer 54.7%.

The key here is the phrase OF ALL MEN or OF ALL WOMEN. That is your clue that you should look within the row % row (or % within respondent's sex row) to find the correct percentage.

Row percentages sum to 100%. So, OF ALL WOMEN, what percentage voted for Clinton = 54.7%. And, OF ALL WOMEN, 12.1% voted for Perot, and OF ALL WOMEN 33.2% voted for Bush. 54.7+12.1+33.2=100% So, when I ask of **ALL WOMEN in the first few words of the question**, I'm asking out of 100% of them – the row %, what % voted for Bush or Perot or Clinton – read the row %.

Column Percentages:

For our data, this is the % within vote for Clinton, Bush, or Perot row. The interpretation of this data is as follows, **OF ALL VOTERS WHO VOTED FOR CLINTON**, 37.1% were men and 62.9% were women. Notice the % sum to 100%, down the column. (See the circles in the table on the previous page.)

Similarly, if I asked you, of **ALL VOTERS WHO VOTED FOR BUSH**, what percentage were women? You should answer 52.3%.

Again, the key here is the phrase OF VOTERS WHO VOTED FOR CLINTON OR BUSH OR PEROT. That is your clue that you should look within the column % row (or % within vote for Clinton, Bush or Perot row) to find the correct percentage.

Chi-Square Statistic (pronounced with a strong ch like kite or like the ch in Scottish loch)

The Chi-square test of independence is used to determine whether the observed values for the cells deviate significantly from the corresponding expected values for those cells. We will use SPSS to compute the chi-square statistic and compare it to the chi-square distribution to see how unlikely the observed value is if the null hypothesis is true.

In order to do the chi-square test, our data must meet the following assumptions:

Assumptions

1) Observations must be independent – individual can only appear once in a table.
   a. Variables can't overlap – for example you could not use age groups like less than 30, 25-40, 35-90 in the same table.
   b. Can't let a person choose 2 favorite car colors and make a table of color preference by gender.
2) Most of the expected counts must be greater than 5 and none less than 1.

Step 4

To generate a Chi-square table in SPSS:

Analyze>Descriptive Statistics>Crosstabs> Row variable=sex, Column Variable = pres92>Click the statistics button and check Chi-square>Continue>ok.

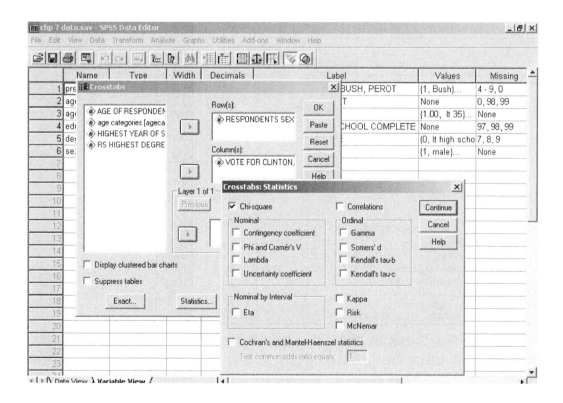

STEP 5:

This is what your output should look like:

Chi-Square Tests

	Value	df	Asymp. Sig. (2-sided)
Pearson Chi-Square	33.830[a]	2	.000
Likelihood Ratio	33.866	2	.000
Linear-by-Linear Association	19.360	1	.000
N of Valid Cases	1847		

a. 0 cells (.0%) have expected count less than 5. The minimum expected count is 121.01.

ASSUMPTIONS

This lets us know if we have met our assumptions. The first thing we check is that most of our expected counts are less than 5. The rule is the % shown in the box should be 20% or less. If it is more than 20%, you must reduce your categories so that most of the expected counts are less than 5. Indeed 0% of our cells have expected counts less than 5, so all of our expected counts in our cells are more than 5. We can look at our cross tab table and see what they are. They were

RESPONDENTS SEX * VOTE FOR CLINTON, BUSH, PEROT Crosstabulation

			VOTE FOR CLINTON, BUSH, PEROT			Total
			Bush	Perot	Clinton	
RESPONDENTS SEX	male	Count	315	152	337	804
		Expected Count	287.7	121.0	395.3	804.0
		% within RESPONDENTS SEX	39.2%	18.9%	41.9%	100.0%
		% within VOTE FOR CLINTON, BUSH, PEROT	47.7%	54.7%	37.1%	43.5%
		Residual	27.3	31.0	-58.3	
	female	Count	346	126	571	1043
		Expected Count	373.3	157.0	512.7	1043.0
		% within RESPONDENTS SEX	33.2%	12.1%	54.7%	100.0%
		% within VOTE FOR CLINTON, BUSH, PEROT	52.3%	45.3%	62.9%	56.5%
		Residual	-27.3	-31.0	58.3	
Total		Count	661	278	908	1847
		Expected Count	661.0	278.0	908.0	1847.0
		% within RESPONDENTS SEX	35.8%	15.1%	49.2%	100.0%
		% within VOTE FOR CLINTON, BUSH, PEROT	100.0%	100.0%	100.0%	100.0%

The second item we check in the Chi-square Tests box is the minimum expected count. It must be greater than 1. Our minimum expected count is 121, which is much higher than 1, so we met that assumption as well.

STEP 6: State chi-square value and p-value.

Chi-Square Tests

	Value	df	Asymp. Sig. (2-sided)
Pearson Chi-Square	33.830[a]	2	.000
Likelihood Ratio	33.866	2	.000
Linear-by-Linear Association	19.360	1	.000
N of Valid Cases	1847		

a. 0 cells (.0%) have expected count less than 5. The minimum expected count is 121.01.

Our **Chi-square value = 33.830.**

Our p-value shows up in SPSS as the number under the column Asymp. Sig. (2-sided) and that number appears to be .000. But in fact, the p-value can never be 0, so we have to double click on our table in SPSS, stretch the right hand side column to the right and see what the p-value really is. Remember, a p-value is a probability so it much be between 0 and 1.

To do this, go to your SPSS output window. Double click on the Chi-square tests table. Once you have double clicked on the table a squiggly line appears around the table, as shown. Go to the far right column, Under Asymp Sig (2-sided), right click and choose Cell Properties, as shown below.

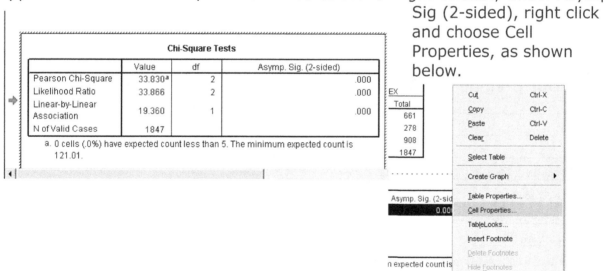

Now, choose the Format Value tab, and choose the 5ᵗʰ option #.## E - ##. This is scientific notation format.

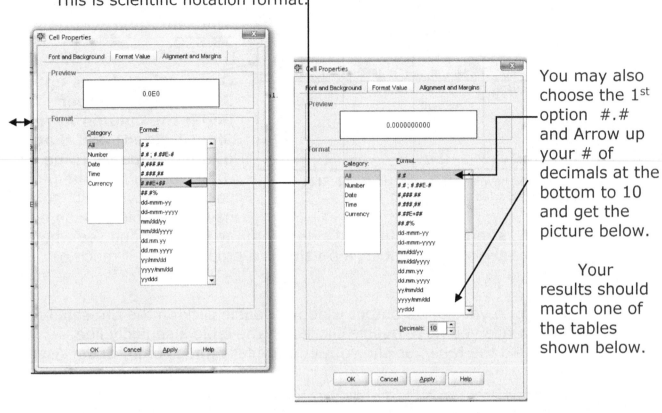

You may also choose the 1ˢᵗ option #.# and Arrow up your # of decimals at the bottom to 10 and get the picture below.

Your results should match one of the tables shown below.

Our p-value = 4.5 ^ -8. This is the same as .00000045.

Chi-Square Tests			
	Value	df	Asymp. Sig. (2-sided)
Pearson Chi-Square	33.830ᵃ	2	0.0000000451
Likelihood Ratio	33.866	2	.000
Linear-by-Linear Association	19.360	1	.000
N of Valid Cases	1847		
a. 0 cells (.0%) have expected count less than 5. The minimum expected count is 121.01.			

Chi-Square Tests

	Value	df	Asymp. Sig. (2-sided)
Pearson Chi-Square	33.830ᵃ	2	4.51E-8
Likelihood Ratio	33.866	2	.000
Linear-by-Linear Association	19.360	1	.000
N of Valid Cases	1847		

a. 0 cells (.0%) have expected count less than 5. The minimum expected count is 121.01.

STEP 7: CONCLUSIONS

Compare your p-value to alpha (.05)

If your p-value is smaller than .05, we reject ho.
If your p-value is larger than .05, we fail to reject ho.

Our p-value is smaller than .05, so we reject ho and conclude there is a relationship between gender and there choice for President in the 1992 election.

If we look at the residuals, we can note that the women have large positive residuals for the response Clinton. This means the observed number of women in those cells is larger than that predicted. Looking at the residuals shows us where the departures from independence are. Women were much more likely to vote for Clinton compared to men. We can also note that women were less likely to vote for Bush and Perot compared to men.

Second example:

"Is education level related to what type of car you drive? Is someone with a High School education more likely to drive an economy rather than luxury car?"

Open the data chisq 2.sav from Blackboard.

Use the variables **CarCat** (Vehicle price category or type) and **Ed** (Level of education).

Step 1:

State your null hypothesis (We write this as H_o:)

Ho: There is **no relationship** between education level and car type.

Step 2:

State your alternative hypothesis (We write this as H$_a$:)

Ha: There is **a relationship** between education level and car type.

Step 3:
Make the cross tabulation table to explore the relationship between our two categorical variables.

Primary vehicle price category * Level of education Crosstabulation

			Level of education					
			Did not complete high school	High school degree	Some college	College degree	Post-undergraduate degree	Total
Primary vehicle price category	Economy	Count	483	587	390	312	69	1841
		Expected Count	399.8	556.9	391.2	389.8	103.3	1841.0
		% within Primary vehicle price category	26.2%	31.9%	21.2%	16.9%	3.7%	100.0%
		% within Level of education	34.7%	30.3%	28.7%	23.0%	19.2%	28.8%
		Residual	83.2	30.1	-1.2	-77.8	-34.3	
	Standard	Count	481	689	485	508	112	2275
		Expected Count	494.1	688.2	483.4	481.7	127.6	2275.0
		% within Primary vehicle price category	21.1%	30.3%	21.3%	22.3%	4.9%	100.0%
		% within Level of education	34.6%	35.6%	35.7%	37.5%	31.2%	35.5%
		Residual	-13.1	.8	1.6	26.3	-15.6	
	Luxury	Count	426	660	485	535	178	2284
		Expected Count	496.1	690.9	485.4	483.6	128.1	2284.0
		% within Primary vehicle price category	18.7%	28.9%	21.2%	23.4%	7.8%	100.0%
		% within Level of education	30.6%	34.1%	35.7%	39.5%	49.6%	35.7%
		Residual	-70.1	-30.9	-.4	51.4	49.9	
Total		Count	1390	1936	1360	1355	359	6400
		Expected Count	1390.0	1936.0	1360.0	1355.0	359.0	6400.0
		% within Primary vehicle price category	21.7%	30.2%	21.2%	21.2%	5.6%	100.0%
		% within Level of education	100.0%	100.0%	100.0%	100.0%	100.0%	100.0%

Answer these questions:

1) Of all the people with Less than a HS diploma, what % drive an economy car?

2) Of all the people with Less than a HS diploma, what % drive a luxury car?

3) Of all the people with a Graduate degree, what % drive an economy car?

4) Of all the people with a Graduate degree, what % drive a luxury car?

5) Can you see a trend when you look at the percentages within level of education across the rows for economy and luxury cars?

6) Of all the people driving economy cars, what % have post graduate degrees?

STEP 5-6: Acquire and state chi-square value and p-value.

Chi-Square Tests

	Value	df	Asymp. Sig. (2-sided)
Pearson Chi-Square	**85.689**[a]	8	**3.48E-15**
Likelihood Ratio	85.520	8	.000
Linear-by-Linear Association	75.149	1	.000
N of Valid Cases	6400		

a.   0 cells (.0%) have expected count less than 5. The minimum
    expected count is 103.27.

STEP 7: CONCLUSIONS

Compare your p-value to alpha (.05)

Our p-value is smaller than .05, so we reject ho and conclude there is a relationship between education level and car choice.

If we look at the residuals, we can note that the people with less than a High School diploma have large positive residuals for the response economy. This means the observed number of people in those cells is larger than that predicted. Looking at the residuals shows us where the departures from independence are. People with less than a high school diploma were much more likely to drive an economy car compared to those with a higher education level. We can also note that people with a post graduate degree were more likely to drive a luxury car.

Chapter 9

Correlation Analysis

Purpose

- To investigate the linear association or relationship between two variables
- In SPSS, the bivariate correlations procedure computes the pair wise associations for a set of variables and displays the results in a matrix.
    - It is useful for determining the strength and direction of the association between two scale or ordinal variables.

We often want to know whether 2 or more variables are related, and if they are, how they are related. We want to know how strong the relationship is and how to make predictions.

A correlation coefficient (r) is a statistic used for measuring the strength of a supposed linear association between two variables. The most common correlation coefficient is the Pearson correlation coefficient. Other types of correlation coefficients are available. Generally, the correlation coefficient varies from -1 to +1.

Example 1 – Use the data file called tipdata.sav found in Blackboard in Course Documents, Data Sets for SPSS Lab book.

Most of us have been confronted with the problem of determining how much tip to leave a waiter or waitress in a restaurant. Although unofficial, many of us have heard the tip should be 15 % of the bill. All of us would believe that there is a relationship between the amount of the bill and the amount of the tip. Consider the sample data from 6 dining parties:

Bill ($)	33.46	50.68	87.92	98.84	63.60	107.34
Tip ($)	5.50	5.00	8.08	17.00	12.00	16.00

Is there sufficient evidence to conclude that there is a relationship between the amount of the bill and the tip?

1. State Null & Alternative Hypotheses

Null Hypothesis: **There is no linear association between the amount of the bill and tip . (r=0)**

Alternative Hypothesis: **There is a linear association between amount of the bill and tip in the population. (r ≠ 0)**

Preparing for Analysis

Create scatter plots to examine the association between the dependent variable and the independent variables

- ○ **Check for presence of linear relationship**
- ○ **Examine strength and direction of relationship**
- ○ **Check for presence of outliers**

Examine correlation coefficients.

Correlation - *A correlation exists between two variables when one of them is related to the other in some way.*

Assumptions:

1. *The sample of paired (x,y) data is a random sample.*
2. *The pairs of (x,y) data have a bivariate normal distribution. This means for any fixed value of x, the corresponding values of y have a distribution that is bell shaped, and for any fixed value of y, the values of x have a distribution that is bell shaped.*
3. *The variables are continuous.*

The second assumption is difficult to check, but a partial check can be made by determining whether the values of both x and y have distributions that are basically bell-shaped, or if we have a small sample size, we can look at Q-Q plots.

Step 2 Explore the data

Make a scatter plot of the data to see the relationship. Open the data on your book CD called tip data.sav.

To make the scatterplot shown in SPSS DO:

Graphs>Legacy Dialogs>Scatter>Simple>tip=dependent variable (y-axis) and Bill = independent variable (x-axis)>ok.

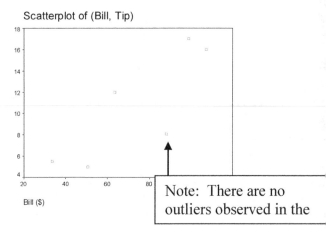

Scatterplot of (Bill, Tip)

Note: There are no outliers observed in the

If the data points make a straight line going from the origin out to high x- and y-values, then the variables are said to have a **positive correlation.** If the line goes from a high-value on the y-axis down to a high-value on the x-axis, the variables have a **negative correlation.**

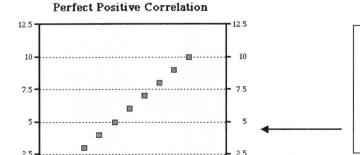

Perfect Positive Correlation

Example – Correlation of total amount of money spent on tickets at the movie theater with the number of people who go. This means that every time that "x" number of people go, "y" amount of money is spent on tickets without variation.

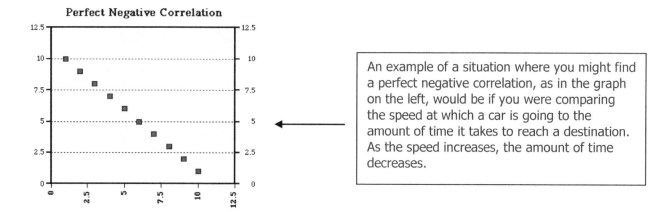

Perfect Negative Correlation

An example of a situation where you might find a perfect negative correlation, as in the graph on the left, would be if you were comparing the speed at which a car is going to the amount of time it takes to reach a destination. As the speed increases, the amount of time decreases.

A perfect positive correlation is given the value of 1. A perfect negative correlation is given the value of -1. If there is absolutely no correlation present the value given is 0. The closer the number is to 1 or -1, the stronger the correlation, or the stronger the relationship between the variables. The closer the number is to 0, the weaker the correlation. So something that seems to kind of correlate in a positive direction might have a value of 0.67, whereas something with an extremely weak negative correlation might have the value -.21.

On the other hand, a situation where you might find a strong but not perfect positive correlation would be if you examined the number of hours students spent studying for an exam versus the grade received. This won't be a perfect correlation because two people could spend the same amount of time studying and get different grades. But in general the rule will hold true that as the amount of time studying increases so does the grade received.

Let's take a look at some examples. The graphs that were shown above each had a perfect correlation, so their values were 1 and -1.

hs below obviously do not have perfect correlations. Which graph would have a
.. of 0? What about 0.7? -0.7? 0.3? -0.3?

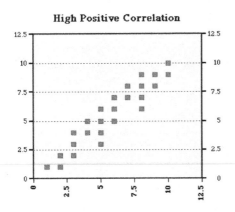

High Positive Correlation

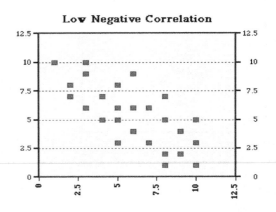

Low Negative Correlation

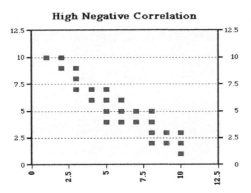

High Negative Correlation

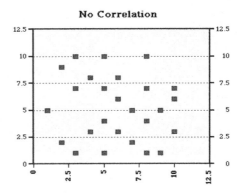

No Correlation

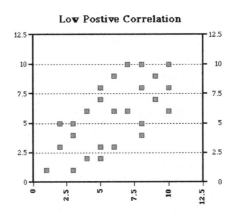

Low Postive Correlation

The linear correlation coefficient r measures the strength of the linear relationship between the paired x and y values in a sample. It is worth mentioning that the statstic r, is not measured on a linear scale. So, an effect with r=0.8 is not twice as big as an effect with r=0.4.

Let's compute the linear correlation coefficient, r, for our sample tip data. Remember – we need to test our assumptions first. The validity of the test requires the data is a random sample and at least one of the variables is normally distributed.

Step 3: Test to see if both variables are normally distributed.

In SPSS, go to Analyze>Descriptive Statistics>Explore

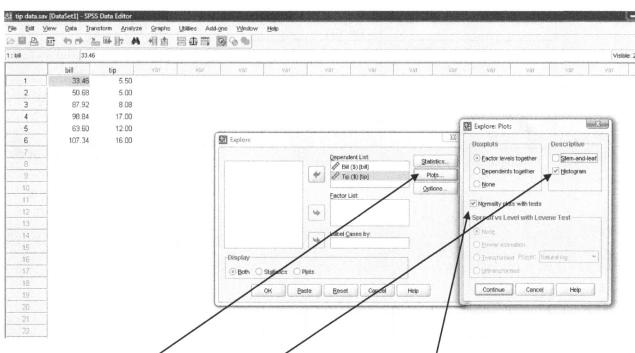

Click the button Plots and check Histogram and Normality Plots with Test> Continue>ok.

The output is shown below.   The Kolmogorov-Smirnov test of normality shows both variables are approximately normal.

Tests of Normality

	Kolmogorov-Smirnov[a]			Shapiro-Wilk		
	Statistic	df	Sig.	Statistic	df	Sig.
Bill ($)	.189	6	.200*	.946	6	.710
Tip ($)	.185	6	.200*	.892	6	.331

*. This is a lower bound of the true significance.

a. Lilliefors Significance Correction

We can also use the Normal Q-Q Plots to see if our data is approximately normal.

reted as follows: Since most of the data points surround the diagonal line, the
° not, we need to use a non-parametric test or transform our data to make it
nd the scope of our class.

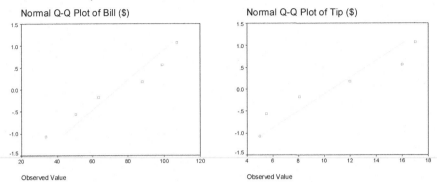

Correlations

		Bill ($)	Tip ($)
Bill ($)	Pearson Correlation	1	.828*
	Sig. (2-tailed)	.	.042
	N	6	6
Tip ($)	Pearson Correlation	.828*	1
	Sig. (2-tailed)	.042	.
	N	6	6

*. Correlation is significant at the 0.05 level (2-tailed).

> To create a correlation in SPSS do this Analyze>Correlate>Bivariate>select bill and tip>Pearson's Correlation Coefficient.
>
> Note: Spearman's and Kendall's are for cases in which the data is not normally distributed.

*Remember - when conducting a formal hypothesis test for co...   No
linear correlation exists and our alternative hypothesis is ther...*

Common Errors Involving Correlation

1. *Concluding that correlation implies causality. – For example, if we showed a strong correlation between the salaries of statistics professors and per capita beer consumption, but these two variables are affected by the state of the economy.*
2. *Using data based on averages. Averages suppress individual variation and may inflate the correlation coefficient.*
3. *The property of linearity – a relationship may exist between x and y even when there is no significant linear correlation. (non-linear or curvilinear association)*

Example 2

Suppose you are interested in finding whether there is an association between people's monthly expenditure and income. To investigate this, you collected data from ten subjects as shown in Table 1 below.

Table 1: Set of paired data

Income / month ($)	Expenditure / month ($)
4000	4000
4000	5000
5000	6000
2000	2000
9000	6000
4000	2000
7000	5000
8000	6000
9000	9000
5000	3000

Step 1:   State the null (ho) and alternative (ha) hypothesis

Ho: There is no linear correlation between income and expenditures.

Ha: There is a linear correlation between income and expenditures.

Step 2: To conduct the correlation analysis, produce a scatterplot of the two variables first.   Open the data **correlation.sav** from Blackboard.

To produce the scatterplot choose:
Graphs >Legacy Dialogs> **Scatter>Simple Scatter**

The **Scatterplot** selection box will be loaded to the screen as shown below, with **Simple** scatterplot selected by default. Click on **Define** to specify the axes of the plot. Enter the variables names *income* and *expenses* into the **y-axis** and the **x-axis** box, respectively. Click on **OK.**

The Scatterplot selection box

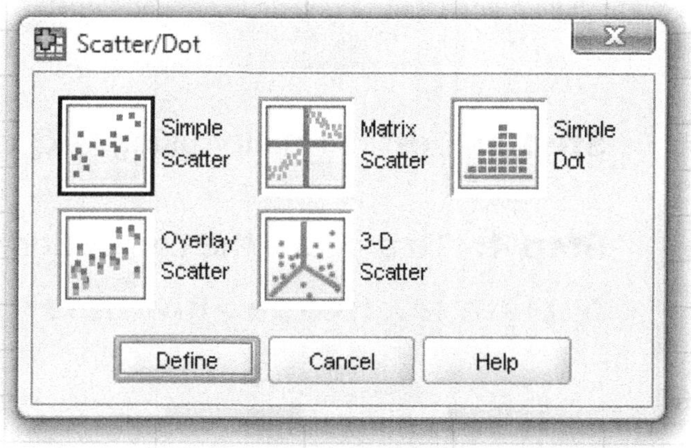

The scatterplot is shown below and it seems to indicate a linear association between the two variables.

Scatterplot Income/month against Expenditure/month

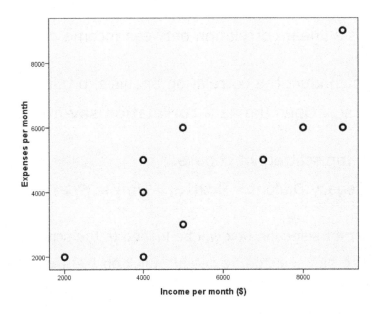

Step 3: Test normality using Q-Q plots or histograms.

Step 4: To produce the correlation analysis choose:

Analyze >Correlate >Bivariate

This will open the **Bivariate Correlation** dialog box as shown below. Transfer the two variables to the Variables text box.

The Bivariate Correlation

dialog box

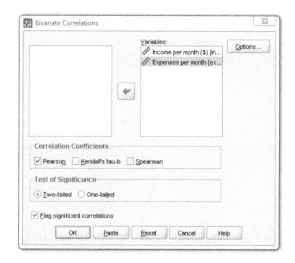

Click on **Options** and the **Bivariate Correlation: Options** dialog box will be loaded on the screen as shown below. Click on the **Means and Standard Deviations** check box. Click on Continue and then **OK** to run the procedure.

The Bivariate Correlation: Options dialog box

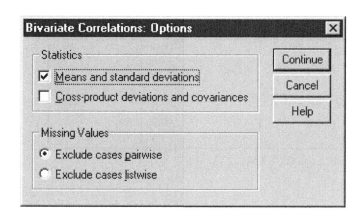

Means and standard deviations are displayed for each variable. The number of cases with nonmissing values is also shown.

Exclude cases pair wise means cases with missing values for one or both of a pair of variables for a correlation coefficient are excluded from the analysis.

Exclude cases listwise means cases or observations with missing values for any variable are excluded from all correlations.

Step 4 and 5: Correlation value, P-value, and Conclusions

Output Listing of Pearson Correlation Analysis

The output listing starts with the means and standard deviation of the two variables as requested under the **Options** dialog box. This result is shown on the table below.

Descriptive Statistics

	Mean	Std. Deviation	N
Expenditure/month	4800.00	2149.94	10
Income/month	5700.00	2406.01	10

The next table from the output listing shown below gives the actual value of the correlation coefficient along with its p-value. **The correlation coefficient is 0.803 and the p-value is 0.005**. From these values, it can be concluded that the correlation coefficient is significant (remember, compare your pvalue to .05). In other words, people with high monthly income are also likely to have a high monthly expenditure budget. There is a significant linear association between income and expenditures per month.

Correlations

		Expenditure/month	Income/month
Expenditure/month	Pearson Correlation	1.000	.803**
	Sig. (2-tailed)	.	.005
	N	10	10
Income/month	Pearson Correlation	.803**	1.000
	Sig. (2-tailed)	.005	.
	N	10	10

** Correlation is significant at the 0.01 level (2-tailed).

Note: 1 asterisk means pvalue < .05
2 asterisks means pvalue < .01

Made in the USA
Coppell, TX
06 September 2020